Manual of Engineering Drawing

Colin H. Simmons
I.Eng, FIED, Mem ASME.

Engineering Standards Consultant, Member of B.S. and I.S.O.
Draughting Committees, Formerly Standards Engineer,
Lucas CAV

Dennis E. Maguire
CEng. MIMechE, Mem ASME, R.Eng.Des, MIED

Design Consultant
Formerly Senior Lecturer, Mechanical and
Production Engineering Department, Southall College
of Technology

A member of the Hodder Headline Group
LONDON • SYDNEY • AUCKLAND

First published in Great Britain 1995 by Edward Arnold
Second impression 1996 by Arnold,
a member of the Hodder Headline Group
338 Euston Road, London NW1 3BH

Whilst the advice and information in this book is believed to be true and
accurate at the date of going to press, neither the author nor the publisher
can accept any legal responsibility for any errors or omissions that may be
made.

British Library Cataloguing in Publication Data
A catalogue record for this book is available from the British Library

ISBN 0 340 58484 X

Typeset in 10/11 Linotron Times by Wearset, Boldon, Tyne and Wear
Printed and bound in Great Britain by J W Arrowsmith Ltd, Bristol

Contents

Preface

This completely revised edition of *A Manual of Engineering Drawing Practice* has been rewritten to include all of the important BS and ISO revisions which are relevant in current draughting practice. During the last twenty years advances in computer technology have gradually made an increasing impact in engineering activities. Methods and procedures in the drawing office, which is the heart of product conception, design and development, have undergone considerable changes with the application of computer power. Drawing office practices involve a vast amount of data manipulation, storage, retrieval and transmission. Stages in the design process require computer models of alternative solutions with associated drawings, calculations and documentation. The ability to speedily incorporate necessary and desirable alterations—to improve solutions as design develops—is vital in competitive industry.

In parallel with computerisation of company draughting facilities has been a firm commitment towards improving product quality. Inferior design can occur from poor company standards or practices, resulting in reduced profits—especially where the product is mass produced, then widely distributed. Typical examples being the recall of cars and electrical equipment for inspections and safety modifications. A growing area of awareness and concern relates to the subject of Product Liability. Litigation resulting from product malfunction often attracts considerable publicity by the media. No company would welcome a loss in stature or reputation due to avoidable design errors or production faults.

A quality product is the result of a sound design followed by adequate checks at each stage in its life, from raw material input to satisfactory operation in service.

Quality requires clearly defined company policies, carefully controlled company procedures and a close adherance to national and, where applicable, appropriate international standards. Quality control aspects of engineering are defined in BS 5750 and start with a clear unambiguous drawing. Although notes on drawings may well be translated into foreign languages, the drawing itself is accepted as a graphical language of communication and its function is to provide an efficient transfer of information between the draughtsman and the end user. This demands that we all speak the same graphic language.

In Great Britain the general intention is to unify in BS 308 one set of standards with recommendations for drawings embracing all areas of mechanical, electrical and electronic engineering associated with the manufacturing industries. However, British Standards in engineering do not operate in isolation but participate actively in International Standards Organization Committees (ISO) assisting in the creation of worldwide agreed standards.

For the benefit of our readers we have included a list of associated engineering standards which will be of assistance to drawing office personnel.

It is important to stress that British and ISO drawing standards are not produced for any particular draughting method. No matter how a drawing is produced, either on an inexpensive drawing board or the latest CAD equipment, the drawing must conform to the same standards and be incapable of misinterpretation.

The text which follows covers the basic aspects of engineering drawing practice required by college and university students also professional drawing office personnel. Applications show how regularly used standards should be applied and interpreted. Geometrical constructions are a necessary part of engineering design and analysis and examples of two and three dimensional geometry are provided. Practice is invaluable, not only as a means of understanding principles, but in developing the ability to visualize shape and form in three dimensions with a high degree of fluency. It is sometimes forgotten that not only does a draughtsman produce original drawings but is also required to read and absorb the content of drawings he receives without ambiguity.

The section on engineering diagrams is included to stimulate and broaden technological interest, further study, and be of value to students engaged on project work. Readers are invited to redraw a selection of the examples given for experience, also to appreciate the necessity for the insertion and meaning of every line. Extra examples with solutions are available in *Teach Yourself Technical Drawing* from TEACH YOURSELF BOOKS.

It is a pleasure to find an increasing number of young ladies joining the staff in drawing offices where they can make an effective and balanced contribution in design decisions. Please accept our apologies for continuing to use the term 'draughtsmen' which is the generally understood collective noun for drawing office personnel, but implies equality in status.

In conclusion, may we wish all readers every success in their studies and careers. We hope they will obtain much satisfaction from employment in the absorbing activities related to creative design and considerable pleasure from the construction and presentation of accurately defined engineering drawings.

Acknowledgements

The authors express their special thanks to the British Standards Institution, 2, Park Street, London, W1A 2BS for kind permission to reprint extracts from their publications.

We are also grateful to the International Organization for Standardization, Genève 20, Switzerland, for granting us permission to use extracts from their publications.

We very much appreciate the encouragement and friendly assistance given to us by:

H C Calton, Ford Motor Company Ltd
Geoff Croysdale, SKF (UK) Ltd
Susan Goddard, KGB Micros Ltd
John Hyde, Norgren Martonair Ltd
Bob Orme, Loctite Holdings Ltd
Tony Warren, Staefa Control System Ltd
Autodesk Ltd

Barber and Colman Ltd
Bauer Springs Ltd
GKN Screws and Fasteners Ltd
Glacier Vandervell Ltd
Lucas Diesel Systems
Lucas Electronic Unit Injector Systems
F S Ratcliffe Ltd
Salterfix Ltd

Dr Russell W H Parry, Senior Editor, and staff of Edward Arnold

Brian and Ray for sheet metal and machine shop examples and models

Our final thanks go to our patient and understanding wives, Audrey and Beryl, for all their typing and clerical assistance since we started work in 1973 on the first edition of *A Manual of Engineering Drawing*.

Chapter 1
Drawing office management and organization

The drawing office is generally regarded as the heart of any manufacturing organization. Products, components, ideas, layouts, or schemes which may be presented by a designer in the form of rough freehand sketches, may be developed stage by stage into working drawings by the draughtsman. There is generally very little constructive work which can be done by other departments within the firm without an approved drawing of some form being available. The drawing is the universal means of communication.

Drawings are made to an accepted standard, and in this country, is BS 308. BS 308 is aligned to the International Organization for Standardization (ISO). These standards are acknowledged and accepted throughout the world.

The contents of the drawing are themselves, where applicable, in agreement with separate standards relating to materials, dimensions, processes, etc. Larger organizations employ standards engineers who ensure that products conform to British and also international standards where necessary. Good design is often the product of teamwork where detailed consideration is given to the aesthetic, economic, ergonomic and technical aspects of a given problem. It is therefore necessary to impose the appropriate standards at the design stage, since all manufacturing instructions originate from this point.

A perfect drawing communicates an exact requirement, or specification, which cannot be misinterpreted and which may form part of a legal contract between supplier and user.

Engineering drawings can be produced to a good professional standard if the following points are observed:

a) the types of lines used must be of uniform thickness and density;
b) eliminate fancy printing, shading and associated artistry;
c) include on the drawing only the information which is required to ensure accurate clear communication;
d) use only standard symbols and abbreviations;
e) ensure that the drawing is correctly dimensioned (adequately but not over-dimensioned) with no unnecessary details.

Remember that care and consideration given to small details make a big contribution towards perfection, but that perfection itself is no small thing. An accurate, well delineated engineering drawing can give the draughtsman responsible considerable pride and job satisfaction.

The field of activity of the draughtsman may involve the use, or an appreciation, of the following topics.

1 *Company communications* Most companies have their own systems which have been developed over a period of time for the following:
   a) internal paperwork,
   b) numbering of drawings and contracts,
   c) coding of parts and assemblies,
   d) production planning for component manufacture,
   e) quality control and inspection,
   f) updating, modification, and reissuing of drawings.
2 *Company standards* Many drawing offices use their own standard methods which arise from satisfactory past experience of a particular product or process. Also, particular styles may be retained for easy identification, e.g. certain prestige cars can be recognised easily since some individual details, in principle, are common to all models.
3 *Standards for dimensioning* Interchangeability and quality are controlled by the application of practical limits, fits and geometrical tolerances.
4 *Material standards* Physical and chemical properties and non-destructive testing methods must be borne in mind. Note must also be taken of preferred sizes, stock sizes, and availability of rod, bar, tube, plate, sheet, nuts, bolts, rivets, etc. and other bought-out items.
5 *Draughting standards and codes of practice* Drawings must conform to accepted standards, but components are sometimes required which in addition must conform to certain local requirements or specific regulations, for example relating to safety when operating in certain environments or conditions. Assemblies may be required to be flameproof, gastight, waterproof, or resistant to corrosive attack, and detailed specifications from the user may be applicable.

6   *Standard parts* are sometimes manufactured in quantity by a company, and are used in several different assemblies. The use of standard parts reduces an unnecessary variety of materials and basically similar components.

7   *Standards for costs* The draughtsman is often required to compare costs where different methods of manufacture are available. A component could possibly be made by forging, by casting, or by fabricating and welding, and a decision as to which method to use must be made. The draughtsman must obviously be well aware of the manufacturing facilities and capacity offered by his own company, the costs involved when different techniques of production are employed, and also an idea of the likely costs when work is sub-contracted to specialist manufacturers, since this alternative often proves an economic proposition.

8   *Data sheets* Tables of sizes, performance graphs, and conversion charts are of considerable assistance to the design draughtsman.

Fig. 1.1 shows the main sources of work flowing into a typical industrial drawing office. The drawing office provides a service to each of these sources of supply, and the work involved can be classified as follows.

1   *Engineering* The engineering departments are engaged on
   a) current production;
   b) development;
   c) research;
   d) manufacturing techniques, which may include a study of metallurgy, heat-treatment, strength of materials and manufacturing processes;
   e) advanced project planning;
   f) field testing of products.

2   *Sales* This department covers all aspects of marketing existing products and market research for future products. The drawing office may receive work in connection with
   a) general arrangement and outline drawings for prospective customers;
   b) illustrations, charts and graphs for technical publications;
   c) modifications to production units to suit customers' particular requirements;
   d) application and installation diagrams;
   e) feasability investigations.

3   *Service* The service department provides a reliable, prompt and efficient after-sales service to the customer. The drawing office receives work associated with
   a) maintenance tools and equipment;
   b) service kits for overhauls;
   c) modifications to production parts resulting from field experience;
   d) service manuals.

4   *Manufacturing units* Briefly, these cover all departments involved in producing the finished end-product. The drawing office must supply charts,

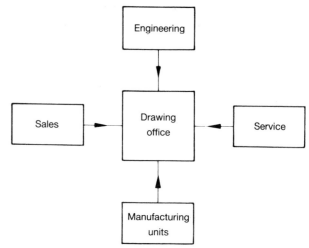

Fig. 1.1

drawings, schedules, etc. as follows:
   a) working drawings of all the company's products;
   b) drawings of jigs and fixtures associated with manufacture;
   c) plant-layout and maintenance drawings;
   d) modification drawings required to aid production;
   e) reissued drawings for updated equipment;
   f) drawings resulting from value analysis and works' suggestions.

Fig. 1.2 shows the organization in a typical drawing office. The function of the chief draughtsman is to take overall control of the services provided by the office. The chief draughtsman receives all work coming into the drawing office, which he examines and distributes to the appropriate section leader. The section leader is responsible for a team of draughtsmen of various grades. When work is completed, the section leader then passes the drawings to the checking section. The standards section scrutinizes the drawings to ensure that the appropriate standards have been incorporated. All schedules, equipment lists and routine clerical work is normally performed by technical clerks. Completed work for approval by the chief draughtsman is returned via the section leader.

Since drawings may be produced manually, or by electronic methods, suitable storage, retrieval and duplication arrangements are necessary. Systems in common use include:

a) filing by hand into cabinets the original master drawings, in numerical order, for individual components or contracts;
b) microfilming and the production of microfiche;
c) computer storage.

The preservation and security of original documents is of paramount importance in industry. It is not normal practice to permit originals to leave the drawing office. A drawing may take a draughtsman several weeks to

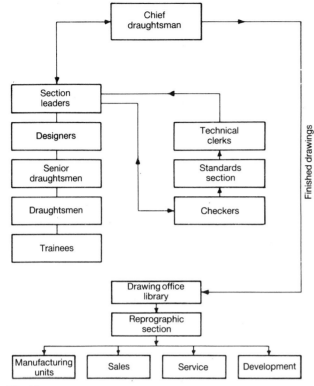

Fig. 1.2

develop and complete and therefore has considerable value. The reprographic staff will distribute copies which are relatively inexpensive for further planning, production and other uses. A library section will maintain and operate whatever archive arrangements are in operation. A large amount of drawing office work comes from continuous product development and modification so easy access to past designs and rapid information retrieval is essential.

Engineering drawing practices

The comments so far refer to drawing offices in general and typical organizational arrangements which are likely to be found within the engineering industry. Good communication by the use of drawings of quality relies on ensuring that they conform to established standards. BS 308 is written in three sections:

Part 1—*Recommendations for general principles*
Part 2—*Dimensioning and tolerancing of size*
Part 3—*Geometrical tolerancing.*

It is perhaps appropriate to clarify the scope of these standards and in the Foreword of Part 1 is the comment

('The recommendations of this standard have been established having regard to the requirements of microfilming. Drawings prepared in accordance with this standard should be suitable for microfilming and reduced-size reproduction techniques.

During the course of this revision particular attention has been paid to developments in computer aided machine draughting and it has been established that such machines are currently capable of complying with this standard. Any minor exceptions due to the limitations of a particular draughting system would not be expected to prejudice the understanding of a drawing otherwise claimed to comply with this standard.')

BS 5070, dealing with engineering diagram drawing practice, is a companion standard to BS 308 and caters for the same industries; it provides recommendations on a wide variety of engineering diagrams. Commonly, as a diagram can be called a 'drawing' and a drawing can be called a 'diagram', it is useful to summarize the difference in the scopes of these standards. BS 308 covers what are commonly accepted to be drawings that define shape, size and form. BS 5070 covers diagrams that are normally associated with flow of some sort, and which relate components (usually indicated by symbols) functionally one to another by the use of lines, but do not depict their shape, size or form; neither may they in general indicate actual connections or locations.

Therefore, any drawing or diagram, whether produced manually or on computer aided draughting equipment, must conform to established standards and will then be of a satisfactory quality for commercial understanding, use and transmission by electronic and microfilming techniques. All of the examples which follow conform to the appropriate standards.

Drawing practice and the computer (CAD: Computer aided draughting and design)

The computer has made a far bigger impact on drawing office practices than just being able to mimic the traditional manual drawing board and tee square technique. However, it depends on drawing office requirements and if only single, small, two dimensional drawings and sketches are occasionally required, then there may be no need for change. CAD can however perform a much more effective role in the design process and many examples of its ability follow—but it will not do the work on its own. The input by the draughtsman needs to follow the same standards applied in the manual method and this fact is often not understood by managers hoping to purchase CAD and obtain immediate answers to design enquiries. The draughtsman needs the same technical appreciation as before plus additional computing skills to use the varied software programs which can be purchased.

To introduce CAD an organization must set out clear

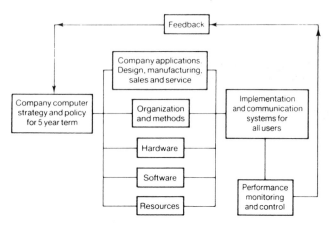

Fig. 1.3   General computer policy relationships

Fig. 1.4

objectives which are appropriate to their present and future requirements and Fig. 1.3 includes aspects of policy which could appear in such plans. The following need consideration:

a) CAD management roles;
b) creation, training and maintenance of capable CAD operators;
c) CAD awareness of design project team members in addition to their leaders;
d) the flow of work through the system and the selecting of suitable types of project;
e) associated documentation;
f) possible changes to production methods;
g) needs involving the customer;
h) system needs relating to planning, security and upgrading;
j) CAD library and database, (Storage of drawings, symbols etc.) and archive procedures.

Many similar aspects will be appropriate in particular applications but good intentions are not sufficient. It is necessary to quantify objectives and provide dates, deadlines, numbers, individual responsibilities and

budgets which are achievable if people are to be stretched and given incentive after full consultation. Present lines of communication will probably need to be modified to accommodate CAD, and planning integration is vital. A possible approach here is the appointment of a CAD Director with the ultimate responsibility for CAD technology assisted by a Systems Manager and an Applications Manager.

A CAD Director has the task of setting and implementing objectives and needs to be in a position to define binding policy and direct financial resources. He will monitor progress. A Systems Manager has the role of managing the computer hardware, the software and the associated data. Company records and designs are its most valuable asset. All aspects of security are the responsibility of the Systems Manager. Security details are dealt with in the next chapter. The Applications Manager is responsible for day to day operations on the CAD system and the steady flow of work through the equipment. He will probably organize training for operators in the necessary computer skills. Both of these Managers need to liaise with the design project leaders to provide and maintain a draughting facility which is capable of increasing productivity to a considerable degree.

Figure 1.4 shows the probable position of the CAD Director in the management structure. His department will be providers of computer services to all other computer users within the company.

Chapter 2
Product development and computer aided design

Work undertaken by a drawing office will vary considerably with different branches of industry. Generally, work of a 'design and make' nature will follow a plan which sets out stages in development from the time a potential client makes an enquiry until the completed product is delivered. The function of the product will dictate many of the associated activities.

A vehicle manufacturer will not design and make all of the parts used but subcontract components from specialists. The engine incorporates electrical and mechanical components and these need to conform to agreed specifications. They must also be designed for installation in specified areas and be suitable for operation in well defined conditions. Component manufacturers strive to improve quality and performance in conjunction with the end user.

The stages in design and development for components in this category are shown typically, step by step, in Fig. 2.1.

1 A client requiring a certain product is often not completely familiar with specific details and needs the experience and advice from a specialist producer to clarify initial ideas. When a range of viable alternatives is presented, opinions can be focused and firm decisions made.

2 The Chief Engineer in a company has the responsibility of producing the company specification for a product. He will no doubt seek advice where aspects of the total design are outside his range of experience, and where design is involved on the fringes of Technology. However a top executive plan needs to be carefully prepared because at the outset the company must know whether or not it wishes to entertain, or get involved with, design proposals to satisfy the client. For example, while rewards may well be great the firm may not be able to cope with the scale of financial and labour demands and delivery requirements in view of current work. They simply may not wish to take the risk and, in view of available production capacity, the firm may prefer not to tender for a possible order.

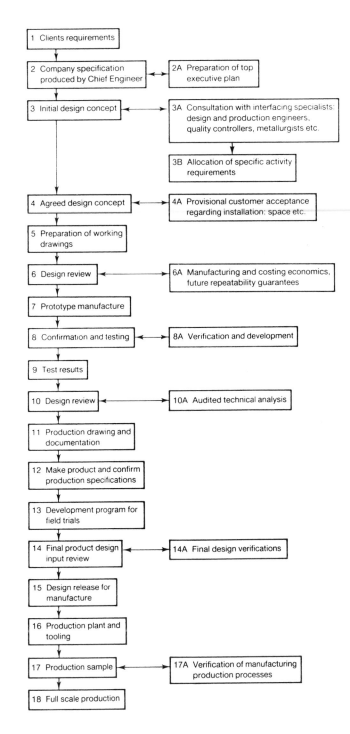

Fig. 2.1

3 Drawings at this stage should be regarded only as provisional. The exercise is needed as an aid to thinking around the problem, with contributions being made by specialists within the firm to ensure feasibility.

CAD has many virtues at this stage of primary design. All information, defined in mathematical terms, can be stored in the system and manipulated on the display. After the basic geometry is established, design variations can be kept and in redrawing alternatives, sections of the previous proposals which were found to be acceptable can be used repeatedly. At any point in development the designer can take a printout, so that suggestions and comments can be made by other technical staff.

It is essential that the Company should appreciate the extent of their commitment if a firm order is accepted at a later date. This commitment includes not only the technical ability to complete the design and manufacture a satisfactory product but also the financial issues relating to its introduction on the factory production line.

4 With the completion of preliminary design work an agreed design concept will have been established, but it is necessary to obtain customer approval before work continues. If our product is to be used in conjunction with others in a large assembly, then, for example, expected overall dimensions and operational parameters need to be confirmed with the client before money is spent on further development.

5 If all is well, working drawings will be prepared. These are not production drawings—at this stage, we as a company have only ensured that our proposals are in line with requirements and that hopefully we shall be able to deliver. The object now is to prepare working drawings to formulate construction methods.

6 A design review is necessary to check the feasibility of manufacturing, to ensure that all aspects of design requirements have been incorporated in an economic manner and to guarantee future supplies.

7 A prototype or a small batch may now be manufactured. The ultimate production methods of manufacture will not be employed here. For example, components which may be moulded could be machined from solid to eliminate casting costs.

8 Prototypes are used for testing to make certain that operational requirements of the specification can be achieved. As a result design changes may be necessary. Product tests cover all areas where the component will be expected to function without failure, and these could include use in extremes of temperature and humidity, also when subject to shock, vibration and fatigue.

9 Proven test results are vital to confirm the validity of these tests.

10 A design review and analysis ensure that progress at this point will be acceptable in every technical aspect to each responsible member of the team.

11 Production drawing can commence now that the performance targets from the prototype have been confirmed. Drawings of the prototype will be reviewed and modifications made to use full scale production processes during manufacture. For plant to be used efficiently plans need to be prepared for loading and progressing work through the factory. The necessary documentation now commences.

12 Manufacture of the final product following production of the prototype has involved modifications and different manufacturing processes. It is therefore prudent to check that the specifications can still be kept.

14 Following trials [13] where the equipment is used in its operational environment and its performance exhaustively checked, the design details can be released [15] for full scale production.

16 Production involves not only the use of machines, but many jigs, fixtures, tools, gauges, inspection procedures need to be planned, and auxiliary equipment designed to move materials on and off production lines.

17 Inevitably teething troubles occur and samples are taken to verify that all plant and equipment operates as planned. Economic production requires that downtime is eliminated before full-scale production commences [18].

Computer aided draughting and design

CAD is much more than drawing lines by electronic means. Similarly, by the purchase of a CAD system, a design does not emerge at the push of a button. 'Buy a computer and you don't need a draughtsman' is also very different from reality. The engineering designer is very much responsible for decisions taken at all technical stages between conception and production. The computer is an aid and performs as it is directed with rapidity and accuracy. The following notes are included to indicate areas of useful activity to assist the draughtsman.

The preparation of two dimensional drawings and the projection of associated views is the 'bread and butter' work in the drawing office. However, draughting three dimensional illustrations is by comparison more time consuming. They are of course used in countless applications since this is how we see the world around us. Service manuals use exploded views so that people with no technical training can follow assembly sequences. Children stick together model kits with guidance using pictorial diagrams.

CAD programs are available where a three dimensional model can be produced automatically given two dimensional views. From the dimensions of the component, the computer will calculate surface areas, volumes, weights for different materials, centres of gravity, moments of inertia and radii of gyration it can also use the applicable values for stress and other calculations, which are a necessary part of design. Computer models permit a study of special relationships and applications are given in the chapter which follows. Models can be manipulated into pleasing forms for artistic approval before production work follows. Previous techniques included modelling with plasticine and plaster, and applications ranged from ornaments to boat hulls and car bodies. CAD has revolutionized modelling capabilities.

Sales departments utilize 3D illustrations in brochures and literature for promotional applications. Desk top publishing from within the company can very simply use illustrations generated as part of the manufacturing process. The scanning of photographs into a CAD system is also an asset especially as photographic work can be retouched, manipulated and animated. Multimedia applications with video and slide presentations form a large part of selling and advertising.

Structural design requires a thorough knowledge of engineering materials properties. Calculations of stress, strain and deflection are essential to determine proportions and dimensions in structural applications. Computers now have the ability to perform millions of calculations per second and with the availability of powerful desk top models, finite element analysis has developed as a principle method. One advantage of finite element analysis is that design engineers can produce better designs and eliminate dubious options during the conceptual design phase. CAD systems permit the rapid generation of models of proposed designs as wire frames. The component can be defined as a collection of small loaded elements. The computer memory stores details of all the geometric data to define each part of the frame. Numerical analysis will then verify whether or not the suggested design will be capable of supporting the expected loads. Formerly, stress calculations were time consuming and in the early days of computing, although the calculation time was considerably shorter, computer time was relatively expensive. This is now not the case and for this type of design work CAD is an essential tool in the drawing office.

CAD is very suitable for repetitive and fast documentation where a product is one in a range of sizes. Assume that we manufacture a range of motor driven pumps operating at different pressures. Many parts will be used in different combinations in the range and the computer database documentation is programmed accordingly. Company standard designs will be offered when enquiries are received. A computerised tender can be sent with the appropriate specification and technical details. On receipt of an order, all of the documentation relating to manufacture, testing, despatch and invoicing will be available. An obvious advantage is the speed of response to the customer's enquiry.

CAD will be linked to CAM (computer aided manufacture) whenever possible. Documentation will include parts lists, materials details of parts to be manufactured or bought out, stock levels, computerised instructions for numerical controlled machine tools, instructions for automated assemblies, welding equipment etc. Printed circuit boards can be designed on CAD and manufactured by CAM.

Production tooling requires the design of many jigs and fixtures. A *jig* is a device which holds the component or is held on to the component, locating the component securely and accurately. Its function is to guide the cutting tool into the component or for marking off or positioning. A *fixture* is similar to a jig but it does not guide the tool. Generally a fixture will be of heavier construction and clamped to the machine tool table where the operation will be performed. Jigs are used frequently in drilling and boring operations. Fixtures are a necessary part of tooling for milling, shaping, grinding, planing and broaching operations. The use of jigs and fixtures enables production to proceed with accuracy, and hence interchangeability due to the maintenance of tolerances (see Chapter 19) and especially by the use of unskilled or semiskilled labour and robotics.

The traditional method of jig and tool draughting was to draw the component in red on the drawing board. The jig or fixture would then be designed around the component. This process ensures that the part is located and clamped correctly, can be loaded and unloaded freely, and that the machining operation can be performed without hindrance.

With a CAD system, the component drawing can be shown in colour on one of the 'layers' (see Chapter 3) and design work undertaken on other layers.

Machining operations need to be checked to ensure that tools and cutters do not foul any other equipment in the vicinity. The path taken by the tool into its cutting position should be the most direct and the shortest in time. The actual cutting operation will take a different time and the tool may traverse the component several times, cutting away more material on each occasion. Machining sequences can be simulated on the screen and when the optimum method has been obtained, the numerical program prepared. All relevant data for the machining operation is converted into coded instructions for continuous production.

Programs are available for the economic use of metallic and non-metallic materials. Many engineering components are manufactured by flame cutting intricate shapes from plate or sheet and these need to be positioned to minimize scrap. The cutting head is guided by computer using the X and Y coordinates at each point along the curve. Other applications use a variety of cutters and saws to shape materials singly or heaped into a pile, such as foams in upholstery or dress fabrics.

The tool draughtsman, for example, will use many standardized components in tooling and designing associated handling equipment for production. If a range of parts is similar it is common practice to produce a single drawing with dimensions in a table of the separate features. A typical example is given in Fig. 7.2 and is the normal manual draughting procedure. CAD can however use a parametric technique where the component drawing is dimensioned by algebraic expressions understood by the computer. Each separate size of component will be given its own part number. When a particular part is required and called up, the computer calculates sizes, draws the part to the correct scale for the draughtsman to position where required on the assembly drawing. This is a very useful facility and only available through the introduction of CAD.

CAD always produces drawings finished to the same high standard, and of a uniform quality and style. All tracing costs are saved.

It will be seen from the above notes that CAD fits in with many of the separate procedures necessary for design and production, but it is vital that, before its introduction, software must be available with proven ability. Likewise, staff must receive training to extract the maximum advantages and benefits.

Draughting in an organization which uses CAD equipment does involve the question of security.

Security aspects of CAD

There are four distinct areas to be addressed:

a) installation and operation;
b) system security;
c) document content;
d) communication.

Installation and operation

Problems can arise from power cuts of short and extended time periods, and from spikes, or fluctuations of power, due to other electrical equipment being switched on. Stormy weather can cause surges and static build up. A reliable power source with a stable supply is essential. Consideration should be given to the provision of a backup supply, if in doubt. Service and maintenance arrangements may require the issue of external contracts, as computer downtime resulting in lost production can prove expensive.

Computers generate heat, and wide variations in environmental temperatures should be avoided. Air conditioning in the complex may be necessary if cooling is required and clean air cannot otherwise be guaranteed. Part of the computer complex may need to be out of bounds except to authorized personnel, to maintain an acceptable environment. Care should be exercised in the selection of floor coverings and furniture to protect equipment from static electricity. Similarly tapes and discs need to be shielded from stray magnetic fields. Ensure that the CAD complex is kept locked and secure when not in use at night and weekends.

Backup copying

An organization must develop a routine for storing the data on which company fortunes may depend. In the event of power failure work in progress may be lost. It could also be lost due to operator error or computer malfunction, fire, flood, vandalism, etc. Backup routines must cover personal responsibility aspects, together with frequency of copying, storage medium and designated places of safety. Backup copies should not be stored in the same buildings as the originals.

System security

Programs used for operating and applying CAD systems need to be checked at regular intervals to ensure that intended methods are being kept in practice.

Document content

Many grades of staff are involved in the design process. Senior designers, detailers, checkers and technical clerks all make a positive contribution. Each member's duties must be carefully defined with rules applied, and authority given, so that each can only operate within his or her agreed sphere of activity. By means of passwords it is possible to access design information at appropriate levels. Revision procedures will ensure that modifications are only made at the correct point by authorized designated staff. Quality assurance systems require strict application of these methods.

Communication

Computer aided designs and production information could easily be copied and some countries do not have legislation prohibiting unauthorized use. Documents should therefore include a clause relating to copyright. Where design information is transmitted, it is recommended that the clause should appear before the text and again at the end.

Access into the computer network

Every CAD installation requires access responsibilities to be defined for the operating staff and the following example relates to an educational establishment.

A typical College of Technology may consist of three separate departments, each requiring to use a common computer facility where a central processing unit is installed. Each department is serviced using a tree and branch system leading to the desks of staff holding different levels of responsibility, and to student outlets in classrooms, drawing offices and laboratories. All members of staff and students need to gain access to the

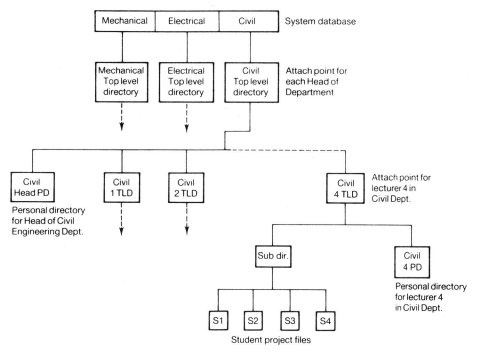

Fig. 2.2 Directory tree for controlled access to database

computer freely, and in their own time, and be able to store their work safely.

A Head of Department, however, may need to gain access to the students' work to monitor progress.

All members of the college staff would wish to have a personal file and keep confidential records. A lecturer must be free to allocate space to students in several classes, so he or she will open subdirectories as necessary and possibly delete work at the completion of a course.

Figure 2.2 shows a directory structure where access can only be made into the system provided the keyboard operator logs in a personal identity number. Each member of staff will be assigned two directories:

a) a top level directory (TLD);
b) a personal directory (PD).

The TLD is the attach point for the user into the system. The lecturer is free to open subdirectories for students' work and each student's file will be protected from the rest of the class. The Head of Department has access to a lecturer's TLD and through to a student's file.

The above system can be adapted for any graded organization where controlled access and protection for records is desirable.

Quality assurance

BS 5750 relates to quality systems and is produced in several sections. The principles of quality assurance embrace all activities and functions concerned with the attainment of quality. *BSI Handbook 22* is essential reading.

Having purchased quality CAD equipment, the products which the company propose to manufacture need to be designed and developed from conception following an agreed *quality assurance* working procedure practised by all employees throughout the organization. QA systems are usually accredited and certified by a third party such as a professional institution or association.

An organization should be able to show that all drawings, documentation and necessary calculations relating to the design, are vigorously checked and approved by management. The stage by stage development of the product will follow an agreed work plan with checks, inspections and correction procedures. Similar plans will cover the manufacturing stages from raw material checks to the tested product. Good communication between all of the participants is essential to ensure that the product meets its specification and the customer's exact requirements.

A company which can demonstrate superior technical skill and expertise has a considerable asset which can be used to advantage in marketing. Proven excellence invariably increases pride and well-being in company employees.

Chapter 3
CAD applications

Rapid development has taken place since the early 1970s when 2D drawing packages were introduced with the ability to draw lines, circles and text. The industry has expanded and progressed and now produces very powerful design and modelling programs. Drawings and notes on the computer screen can be modified and edited with changes displayed in an instant.

A wide range of screen sizes are available in monochrome or colour. Any shade of colour is obtainable by the graphic artist. Drawings are reproduced in many sizes and small items present little difficulty with zoom facilities. Views drawn to different scales can be arranged on the same drawing print.

Flexible drawing systems provide a drawing area and a menu on the screen. 'Help' advice is freely available at the press of a button. Selecting and drawing from the menu is done by using a device such as a mouse or digitiser. Choosing from the options listed on the screen is performed by pointing and clicking the device.

The provision of multi layers provides a very useful method of working on CAD. Imagine transparent sheets placed on top of each other which may be shuffled and rearranged so that you can draw on the top. Each of the layers underneath in the pile can be turned on or off, they may be given identification colours and selected parts of drawings moved from layer to layer if required.

Assume that we want to draw plans for a house. Layer 1 could be used to draw a plan view of the building plot. Layout work is often easier if graph paper is used. On layer 2 we make our own construction grid which is transparent graph paper with squares to any convenient scale of our choice. Using this grid under layer 3 we design a suitable ground floor layout. Layer 4 showing the first floor layout could be started by tracing the position of the outside walls from layer 3. When all of the required plans and elevations are constructed, they can be repositioned on a drawing arrangement; and, if necessary, the site layout reduced to a smaller scale. When completed, the construction grid may be deleted. Tracing facilities and the ability to print layers together or apart are a valuable draughting asset.

The physical equipment components of a computer system are known as the *hardware*. The programs and data used on the computer are defined as the *software*.

Another advantage of CAD is its ability to store line systems and other entities which are frequently used on drawings. For example, software containing symbols to British, European and other International Standards is freely available for most engineering applications. Libraries of regularly used parts can be created by the draughtsman. For repetitive use on a drawing, a typical item may be retrieved and positioned in seconds, also oriented at any angle to suit particular circumstances.

As a drawing aid, every CAD program must provide geometric features, permitting the operator to blend lines and arcs. It is necessary in engineering drawing to be able to determine points of tangency between straight lines and curves and between curves of different radii. Productivity is much improved by a program enabling you to easily draw polygons, ellipses, multiple parallel lines and multiple parallel curves. The speed of machine drawing is increased by the use of automatic fillets and chamfers. Layout work benefits when use is made of construction grids; and the computer's ability to 'snap' automatically to particular geometric points and features, will speed the accurate positioning of linework. Copy, rotate and mirror facilities are also very handy when drawing symmetrical parts.

Automatic cross hatching within closed boundaries is useful in the construction of sectional views and when indicating adjacent parts and different materials. Many changes of hatch patterns are supplied with CAD programs. Filling areas in various colours is a requirement in artwork.

The ability to zoom in and out is an asset when drawing to scale. It is possible to work on fine detail in an assembly and then zoom out to observe the result in context.

CAD information is stored in digital form and hence, irrespective of the size of the final printed drawing, it is possible to accurately dimension components automatically.

Different 'type-set' and alternative style fonts are always supplied with CAD programs. If a special font is required to match an existing style then specialist vendors can supply. Alphabets in different languages present no problem.

Quite clearly the physically largest affordable screen has many advantages. If the draughtsman is involved with desk top publishing, it is ideal to be able to work on a screen which displays two A4 paper sheets side by side so that 'what you see is what you get'. The screen should give high resolution (1024×768 pixels minimum); this is necessary to give an image which is flicker free. The quality of the display will have a big contribution to make in the avoidance of fatigue and eye strain. On some installations the graphics are drawn on one high resolution screen with documentation and calculations directed to a smaller display. First hand practical experience is important here for an ideal solution.

Plotting and printing equipment will vary according to drawing office requirements. It is true however that many CAD installations are judged by the quality of their plotted drawing. It is necessary to have a demonstration and this will ensure that an excellent CAD system will have an output to do it justice.

A wide variety of plotters are available for reproductions from A4 to A0 in size, in colour or monochrome, and in a quality suitable for production work or the most prestigious presentations.

Autodesk Ltd (Cross Lanes, Guildford, Surrey GU1 1UJ) is the world's leading producer of CAD visualisation and animation software for personal computers and workstations. AutoCAD programs have been adopted in educational establishments, and since 1987 certified national courses of study by the City and Guilds of London Institute have been conducted throughout the country. Authorised training centres cater for the needs of local industry and of those who wish to develop their CAD skills further.

Autodesk has been at the forefront of applying standards within the computer aided design environment.

The main professional program AutoCAD is very much a non-specific or generic CAD tool and many applications are available to the basic graphics package which enhance its suitability for a particular discipline. The *AutoCAD Applications Handbook* which is a CAD User Publication lists more than 500 software packages which can be used to maximize productivity in association with AutoCAD.

AutoSketch

A typical starter CAD program is AutoSketch which is easy both to learn and to use. It is a low cost package, ideal for anyone who wants to use a computer to sketch or draw without investing in a full-scale system.

Drawings are created by choosing drawing and editing commands from pull down menus. Drawings, patterns and fonts are represented by simple symbols, or icons. You can draw on multiple layers and look at them in any combination. Repetitive drawing is eliminated; you can use previously created drawings to build libraries of frequently used symbols, saving time.

Having drawn an object you can move, copy, rotate, mirror, stretch and erase it until it matches your needs. You can group components together and treat them as one, and break them apart for editing. The UNDO command permits drawing changes, or to change back again, use REDO. Expanded memory support allows you to work with large drawings. Part clipping allows you to select items from existing drawing files and use

them in others. Text and notes can easily be added or edited on the drawing using a variety of fonts. The drawing features include line options, arcs, ellipses, circles, points, pattern fill areas, spline curves and polylines with variable line width. Automatic fillets and chamfers are possible, and the program also offers zoom and pan facilities.

The program uses the industry standard DXF format, which allows you to export drawings directly into AutoCAD, and a wide selection of desktop publishing packages.

Computer requirements: IBM PC/XT/AT or PS/2 or 100% compatible. 640K RAM. A hard disk and floppy drive.

AutoCAD AEC

A comprehensive 2D and 3D draughting and design software package is available for architectural, engineering and construction work. From a 2D site survey, as an Ordnance Survey DXF file for example, you are able to start original design schemes. Speed and ease of revision permit different ideas to be considered without the need to redraw the entire plan. If the site survey is in 3D, you can also create 3D models and incorporate them into the site. An automatic perspective facility will allow you to view the design from any viewpoint, and to check on the look and feasibility of the project. Animated perspective walkthroughs can be produced with ease. BS 1192 Part 5, specifies a layered format and the scheme design could assemble 2D general arrangement drawings and layouts.

A vast library of architectural symbols is at your disposal. To insert a door into a wall, for example, you select one from a large range of door types, point to where you want to place it, the program will break the wall, insert the door and form the relevant detail. If you change your mind, repositioning is easy: just move it and the previous wall opening will automatically readjust itself. Bearing in mind that the library can contain all the standard types of doors, windows and furnishing items etc. that you regularly use or supply, then a proposal drawing is easy and quick to produce. At this stage initial costings and specifications can also be prepared. Facilities are also available for the design and modelling of the roof and staircases.

At the production stage a 3D model can be assembled from the approved detail available from 2D floor layouts, elevations and sectional details. The management facility in the program allows you to specify and attach production information to the drawings, and to extract it for estimating purposes and for the preparation of bills of material quantities. All members of a project team can share accurate and detailed information. Adapted drawings can be used by facilities managers to plan or change the function of a building for the benefit of occupants.

AutoCAD AME

Solid modelling uses a computer to develop a mathematical description of three dimensional objects and shapes. The description must contain all the information necessary for engineers to visualize, analyse and simulate the manufacturing process—from a single component to a complete assembly. The description includes information about edges, surfaces, material properties, volumes and relationships with other objects. The information must be accurate and unambiguous.

Engineers can create objects from standard solid shapes as well as user-defined shapes generated by extruding or revolving profiles. The process is known as a Boolean operation and a 3D model can be formed by a combination of these methods:

a) the addition of any two 3D objects or shapes;
b) the subtraction of any two 3D objects or shapes;
c) using the volume common to the two 3D objects or shapes.

The computer remembers the sequence in which the operation was carried out but cannot handle what might be described as a freeform shape.

In another modelling system the computer maintains a detailed description of edges and surfaces.

This is a program which combines the advantages of both methods.

A particularly useful application of solid modelling is in plant engineering. Pipework layouts can be constructed from plans and elevations, and checks made to ensure that adequate clearance exists between adjacent pipe services. Models are also valuable for teaching plant operation and safe practices.

Stress calculations for different materials are a necessary part of engineering design and modelling programs provide facilities for the extraction of areas, volumes, weights, centres of gravity etc.

Autodesk 3D Studio

This is available to the designer who needs visualization to promote and sell ideas. The range of applications is extensive covering packaging, furniture design, architecture, business presentations, stage set designs, factory and process simulation, industrial and consumer product design and robotic simulation.

The program runs on industry standard PCs and supports a wide range of input and output devices.

AutoVision

AutoVision is an additional software package which can be used with AutoCAD Release 12 to provide high-quality photorealistic still renderings. These enhance the appearance of models and presentations.

Design concepts which are rendered, clearly and convincingly, certainly aid at the stage where decisions need to be made to finalize aspects of shape, form and finish. The presentation of alternative solutions using the same master drawing is also an added bonus.

Engineered components are often designed for clients without a technical background. To be able to observe the final product in three dimensions with its approved finish and in an ideal situation will reduce design time. Many people have difficulty in reading drawings, but with a presentation of an internal building detail which perhaps shows a slate floor and coloured textured walls, then the client can understand exactly how the structure will look. The drawing bridges the communication gap.

Creating renderings is fast and since the data required is stored in the DWG file it cannot get separated from the component drawing. Menus and dialogue boxes are used. The program features include shading and ray tracing giving shadowing, reflection and refraction effects. A comprehensive library of materials and textures can be used to create a variety of surfaces, such as wood, glass, marble, granite etc.

A wide selection of illumination tools and compatibility with 3D Studio Release 3 software, will allow the draughtsman to make walk-throughs, fly-throughs and animated product-assembly presentations.

Lighting studies are easy and accurate. You can produce a variety of artificial, natural and mixed lighting effects. It is possible to arrange directional lights in various combinations and locations and control such characteristics as colour, intensity, attenuation and shadowing.

In addition, a Sun Locator lets you work easily with sunlight effects. You can position the sun to a specific time of day and year to create realistic sun-shadow combinations. The feature allows architects to calculate, for example, whether a living room will receive enough sunlight at midday in late December. On a larger scale, in the design of shopping malls for example, the position of the sun in relation to a particular area can materially effect heating, lighting and cooling loads.

You will appreciate that these programs can help to detect design mistakes or misunderstandings while they are still easy and inexpensive to remedy.

Typical AutoCAD drawings

The following examples are meant to convey to the reader the extensive range of draughting facilities available from software associated with the basic AutoCAD Release 12 program. Obviously there is a certain amount of overlap in the scope of programs and many ways of performing the same operation. It is recommended that advice should be taken and a demonstration requested for a special type of application.

Fig. 3.1 shows three pictorial drawings of an engine development.

Orthographic drawings of a cycle have been produced by AutoCAD in Fig. 3.2. The three dimensional illustration has also been rendered using AutoVision.

Fig. 3.2

Fig. 3.3 BS4006 gives the dimensions, testing and design requirements for hand-operated square drive socket wrenches and accessories. The tools are manufactured from chrome vanadium steel and these illustrations show typical production drawings and presentations for sales brochures.

Fig. 3.4

Architectural drawings for the design of a shopping mall are reproduced in Fig. 3.4. They show the outline of the development and how the complete construction could appear.

Fig. 3.5

Fig. 3.5 shows an architectural drawing and demonstrates the realistic effect of rendering. Computer simulation may be used to assist the client in the choice of colour for the finished building. Alternative solutions are easily prepared.

Fig. 3.6

A perspective drawing of an internal part of a building in Fig. 3.6 indicates the style and character of a finished construction.

Fig. 3.7

Fig. 3.8   Examples of animated presentations are given in
Fig. 3.7 and Fig. 3.8. Basic engineering constructions
are often enhanced with theatrical effects in sales
literature.

Fig. 3.9 illustrates the variety of subtle textures available within the materials library

The manufacture of many domestic appliances, for example, requires moulds to be made in which components can be produced from a variety of plastics. In Fig. 3.9 the path taken by the tool, which will machine the mould, can be simulated to ensure product accuracy. The detail shows part of a production drawing for a hairdryer mould.

AutoCAD LT

For many CAD users AutoCAD is too comprehensive, advanced and expensive for their needs and in cases where the draughtsman is mainly responsible for layout work and design work, which does not involve sophisticated modelling and rendering, then this program is well worth considering. The current price is less than £500.

The system demands a 386, 486 or Pentium-based PC with co-processor support, MS Windows 3.1, 4MB of RAM and at least 10MB hard disk space. This will also accommodate a selection of symbol files.

The program contains powerful 2D and basic 3D geometry creation, editing, display and plotting options and uses the AutoCAD DWG file format so interchange of existing files presents no problem.

The ability to draw and to use the drawing, or part of the drawing, in a word processor document will be much appreciated. The software has a familiar feel as it uses the Windows interface, so if you have used pull-down menus, dialogue boxes and the drag and drop simplicity of Windows, you will soon master the basics since the Help feature provides an on-line guided program tour. The 2D draughting content is identical with that of Release 12 of AutoCAD. The toolbox and toolbar can easily be customised and quickly rearranged to suit your own style of working.

Another feature of value is a window which gives you an overall view of your drawing for fast zooming and panning. It is possible to link drawings so that if you update the master, linked drawings are updated automatically. Modifications use quite a large proportion of drawing office time.

Fig. 3.10   illustrates exploded three dimensional views of a turbocharger for an automobile.

Fig. 3.11   shows another application of AutoCAD LT where part of an assembly has been copied into a word-processor document and used to prepare production engineering information.

Project development

The reader will appreciate that the design of, for example, a large construction project from its conception, will involve technical input from architects and engineering designers in a wide variety of associated disciplines. It is vital that all contributors to the overall scheme talk the same language and that only compatible computer software packages are in use for the separate areas of work. In addition, the management contractor must have access to the designs as work is in progress. Before the age of CAD it was the practice to have countless meetings in order to coordinate progress. Design obviously continues in steps and in planning and construction work problems arise, and designers need to be in a position to make modifications to overcome them, before progressing to the next phase.

A typical case study which illustrates the activity associated with this type of work is in the construction of the new Civil Aviation Authority 'en-route' centre, being built at Southampton. This prestige building and installation will control all the air traffic passing through Britain's airspace and house controllers operating banks of electronic and computer equipment where only an efficiency of 100% is acceptable. The building services engineer must also ensure that the environment to keep both controllers and equipment comfortable is maintained 24 hours a day, seven days a week.

Due to the extensive use of computers at the centre, a huge amount of electrical, heating, ventilating and air conditioning plant needs to be installed. Different specialist contractors are responsible for these services under the stewardship of the management contractor.

The fast track nature of the design and construction, requires an extensive application of CAD, where individual contractors responsible for electrical, mechanical and ducting work, are 'net-worked' on site, and can call up CAD data from each other.

At this development, it was accepted by contractors that for some drawings it would be practical to work in three dimensions to make it easier, for example, to ensure clearances between piping and ductwork in the more cramped areas. Layout drawings in 3D permit an engineer to demonstrate clearly to another party where, for example, an electrical cable is likely to plough straight through a duct. Potential problems are solved on screen rather than emerging during construction. In addition, access for maintenance purposes and replacement of equipment is a vital consideration.

In the design of many heavy engineering plant layouts it was the practice to build a scale model of the plant as design work progressed. The function of the model was to check on the feasibility of the installation. Any obvious improvements could then be incorporated.

Typical examples would be during the design of chemical plants and oil refineries. After the completion of the project, the model was used to assist in the education of technicians who would later operate and service the equipment.

Drawings in 3D can materially assist in comprehension, and the draughtsman can check designs by altering the angles from which arrangements are viewed on screen.

When many work stations have to be installed for a design team, it is vital to agree on working methods. BS1192 has been developed for Construction Drawing Practice and the contents of the various sections are detailed in Chapter 27. Agreement is necessary on the organisation of many aspects of work and in CAD, these include the use of layers, the groupings of the various sections of construction designs, use of colours so that similar ductwork appears on the screen in the same shade, procedures for the transfer of data between several drawing offices, methods of structuring data for archiving and to help future retrieval. Text fonts need to be agreed so that drawing quality is uniform. All drawing work must also be to BS308 for a complete understanding and to avoid ambiguity. It is essential that all contributors work as a team and in harmony if planning deadlines are to be kept, as obviously, delays in one area of construction can hold up another contractors work, and may result in financial loss.

The designs for services and installations originate from specifications and schematic layouts supplied by Consulting Engineers acting on behalf of the Clients or Agents.

For layout work, a draughting package which covers all aspects of services, such as, Electrical, Lighting, Communication, Alarms, Ductwork, Sanitary and Mechanical plant, all covered by BS1192, is desirable and time saving. Standard symbols can be inserted on their appropriate drawing layer, rotated automatically to align with a wall or ceiling grid and automatically scaled so that they are plotted at the correct scale. These settings can also be customized, to enable you to predefine commonly used layers and sizes.

The Building Services Library supplied by;

HEVACOMP,
109 Regents Park Road,
London
NW1 8UR.

is a typical package which covers these requirements and permits you to store up to 600 of your own symbols using tablet or pull down menus. The package will assist in the creation of working drawings and in the detailing of, for example, sections of ductwork, the program will prompt for the dimensions, elevation and layer; subsequent sections of ductwork are then able to attach automatically, matching the layer and size.

Parametric routines are also used to efficiently design a wide variety of bends, tees, branches and transition pieces for all types of square, rectangular, circular and oval ductwork.

Schedules of fittings need to be created with essential

information and if necessary interfaced with other database or spreadsheet programs, in order to prepare bills of materials.

Electrical wiring systems for lighting and services must be designed in accordance with I.E.E. Wiring Regulations and programs are available to provide the essential requirements of both the electrical designer and contractor.

The ELEC program from HEVACOMP can be used to calculate all cable, cpc and fuse sizes, as well as voltage drops, earth fault loop impedances, and short circuit currents. Schematic diagrams are easily prepared up to A0 in size, showing load descriptions, protective device and cable sizes as well as sub-main details.

Reporting facilities include:
Board and way data,
Cable sizes and voltage drops,
Short circuit currents and disconnection times,
Discrimination charts,
Input data,
Calculation file data,
Load current summary,
Voltage drop summary,
Equipment schedules.

Size of computer

As a rough guide to selection, the larger the drawing and degree of complexity, the more important is the performance of the machine. If a drawing contains large areas which are cross-hatched or shaded it is important to be able to redraw illustrations quickly to prevent time wasting.

It is easy to obtain demonstrations of computer power and this is recommended.

Generally for small to medium sized drawings, an 80386 machine with a maths coprocessor running at 25 MHz with a minimum of 8MB of internal memory (RAM) may be sufficient. Larger drawings may need a 80486 machine running at 25/33 MHz with not less than 8 MB of RAM. For solid modelling where colour rendering is a functional requirement, an 80486 running at 33 MHz with not less than 16 MB of RAM would be more suitable. In addition the computer should have a hard disk of 60 MB minimum.

Future trends are for applications needing greater amounts of computer memory so the chosen system should be expandable.

When selecting software products which are required to operate with each other it is necessary to check compatibility; your dealer should advise.

Fig. 3.7 An assembly drawing of a fuel injector for a diesel engine. Drawn to BS and ISO standards, this is a typical professional CAD drawing which could be produced using AutoCAD LT or software with a similar specification.

Chapter 4
Principles of first and third angle orthographic projection

First angle projection

Assume that a small block is made 35 mm × 30 mm × 20 mm and that two of the corners are cut away as shown below in three stages.

Fig. 4.2 illustrates a pictorial view of the block and this has been arranged in an arbitrary way because none of the faces are more important than the others. In order to describe the orthographic views we need to select a principal view and in this case we have chosen the view in direction of arrow A to be the view from the front.

The five arrows point to different surfaces of the block and five views will result. The arrows themselves are positioned square to the surfaces, that is at 90° to the surfaces and they are also at 90°, or multiples of 90°, to each other. The views are designated as follows:
View in direction A is the view from the front,
View in direction B is the view from the left,
View in direction C is the view from the right,
View in direction D is the view from above,
View in direction E is the view from below.

In first angle projection the views in the directions of arrows B, C, D and E are arranged with reference to the front view as follows:
The view from B is placed on the right,
The view from C is placed on the left,
The view from D is placed underneath,
The view from E is placed above.

The experienced draughtsman will commit the above rules to memory. It is customary to state the projection used on orthographic drawings to remove all doubt, or use the distinguishing symbol which is shown on the arrangement in Fig. 4.3.

Fig. 4.1

Fig. 4.2

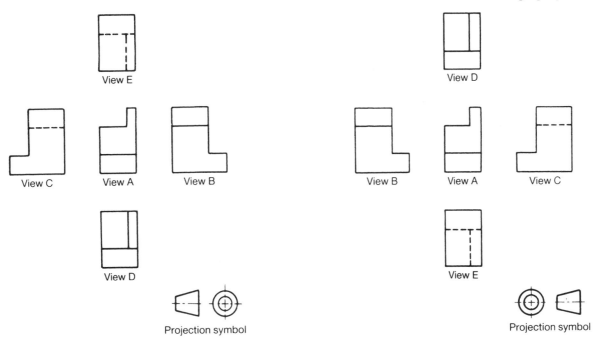

Fig. 4.3 First angle projection arrangement. Dotted lines indicate hidden edges and corners

Fig. 4.4 Third angle projection arrangement

Third angle projection

The difference between first and third angle projection is in the arrangement of views and, with reference to the illustration in Fig. 4.4, views are now positioned as follows:

View B from the left is placed on the left,
View C from the right is placed on the right,
View D from above is placed above,
View E from below is placed underneath.

Study the rearrangement shown below in Fig. 4.4 and remember the above rules because it is vital that the principles of first and third angle projection are understood. The distinguishing symbol for this method is also shown.

If a model is made of the block in Fig. 4.1, and this can easily be cut from polystyrene foam used in packing, then a simple demonstration of first and third angle projection can be arranged by placing the block on the drawing board and moving it in the direction of the four chain dotted lines terminating in arrows in Fig. 4.5. Fig. 4.5(a) shows the positioning for first angle and Fig. 4.5(b) for third angle projection. The view in each case in the direction of the large arrow will give the five views already explained.

The terms first and third angle correspond with the notation used in mathematics for the quadrants of a circle. In Fig. 4.6 the block is shown pictorially in the first quadrant with three of the surfaces on which views are projected. The surfaces are known as planes and the principal view in direction of arrow A is projected on to the principal vertical plane. The view from D is pro-

jected on to a horizontal plane. View B is also projected on to a vertical plane at 90° to the principal vertical plane and the horizontal plane and this is known as an auxiliary vertical plane. Another horizontal plane can be positioned above for the projection from arrow E, also a second auxiliary vertical plane on the left for the projection of view C. Notice that the projections to each of the planes are all parallel, meeting the planes at right angles and this is a feature of orthographic projection.

The intersection of the vertical and horizontal planes give a line which is the ground line GL. This line is often referred to as the *XY* line; this is useful in projection problems since it represents the position of the horizontal plane with reference to a front view and also the position of the vertical plane with reference to a plan view. Many examples follow in the text.

If the planes containing the three views are folded back into the plane of the drawing board, then the result is shown in Fig. 4.7 where dimensions have also been added. The draughtsman adjusts the distances between views to provide adequate spaces for the dimensions and notes.

To describe a simple object, a draughtsman does not need to draw all five views and it is customary to draw only the minimum number which completely illustrate the component. You will note in this particular case that we have omitted views which contain dotted lines in preference to those where corners and edges face the observer. Many parts do not have a definite 'front', 'top' or 'side' and the orientation is decided by the draughtsman, who selects views to give the maximum visual information.

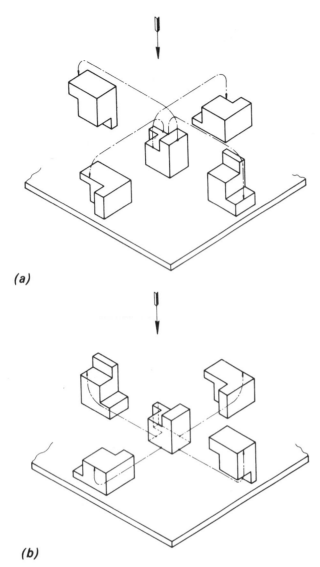

(a)

(b)

Fig. 4.5   (a) First angle arrangement
(b) Third angle arrangement

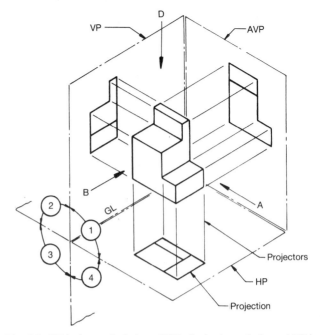

Fig. 4.6   VP is the vertical plane. HP is the horizontal plane. AVP is the auxiliary vertical plane. GL is the ground line

Fig. 4.7

Traditionally, front views are also known as front elevations, side views are often known as side or end elevations and the views from above or beneath are referred to as plans. All of these terms are freely used in industrial drawing offices.

Projection symbols

First angle projection is widely used throughout all parts of Europe and often called European projection. Third angle is the system used in North America and alternatively described as American projection. In the British Isles, where industry works in co-operation with the rest of the world, both systems of projection are regularly in use. The current British and ISO standards state that both systems of projection are equally acceptable but they should never be mixed on the same drawing. The projection symbol must be added to the completed drawing to indicate which system has been used.

Figure 4.8 shows the recommended proportions of the two projection symbols.

Fig. 4.9 indicates how the First angle symbol was obtained from projections of a tapered roller. The Third angle alternative is given in Fig. 4.10.

Please note the movement suggested by the arrows in Fig. 4.9, Fig. 4.10 and also in Fig. 4.8, since orientation is the main clue to understanding the fundamental differences in projection systems.

An experienced draughtsman must be fully conversant with all forms of orthographic and pictorial projection and be able to produce a drawing where no doubt or ambiguity relating to its interpretation can exist.

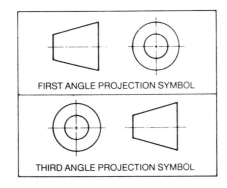

FIRST ANGLE PROJECTION SYMBOL

THIRD ANGLE PROJECTION SYMBOL

Fig. 4.8

Fig. 4.9

Fig. 4.10

Drawing procedure

Generally, industrial draughtsmen do not complete one view on a drawing before starting the next, but rather work on all views together. While projecting features between views, a certain amount of mental checking takes place regarding shape and form, and this assists accuracy. The following series of drawings shows stages in producing a typical working drawing in first angle projection.

Stage 1 (Fig. 4.11). Estimate the space required for each of the views from the overall dimensions in each plane, and position the views on the available drawing sheet so that the spaces between the three drawings are roughly the same.

Stage 2 (Fig. 4.12). In each view, mark out the main centre-lines. Position any complete circles, in any view, and line them in from the start, if possible. Here complete circles exist only in the plan view. The heights of the cylindrical features are now measured in the front view and are projected over to the end view.

Stage 3 (Fig. 4.13). Complete the plan view and project up into the front view the sides of the cylindrical parts.

Stage 4 (Fig. 4.14). Complete the front and end views. Add dimensions, and check that the drawing (mental check) can be redrawn from the dimensions given; otherwise the dimensioning is incomplete. Add the title and any necessary notes.

It is generally advisable to mark out the same feature in as many views as is possible at the same time. Not only is this practice time-saving, but a continuous check on the correct projection between each view is possible, as the draughtsman then tends naturally to think in the three dimensions of length, breadth and depth. It is rarely advantageous to complete one view before starting the others.

Reading engineering drawings

The following notes and illustrations are intended to assist in reading and understanding simple drawings. In all orthographic drawings, it is necessary to project at least two views of a three dimensional object–or one view and an adequate description in some simple cases, a typical example being the drawing of a ball for a

Fig. 4.11 Stage 1

Fig. 4.12 Stage 2

Fig. 4.13 Stage 3

Fig. 4.14 Stage 4

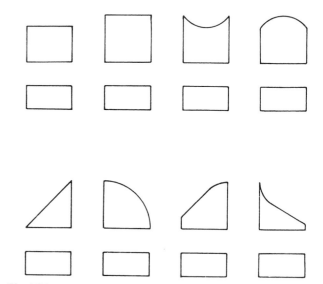

Fig. 4.15

bearing. A drawing of a circle on its own could be interpreted as the end elevation of a cylinder or a sphere. A drawing of a rectangle could be understood as part of a bar of rectangular cross-section, or it might be the front elevation of a cylinder. It is therefore generally necessary to produce at least two views, and these must be read together for a complete understanding. Fig. 4.15 shows various examples where the plan views are identical and the elevations are all different.

A single line may represent an edge or the change in direction of a surface, and which it is will be determined only by reading both views simultaneously. Fig. 4.16 shows other cases where the elevations are similar but the plan views are considerably different.

A certain amount of imagination is therefore required when interpreting engineering drawings. Obviously, with an object of greater complexity, the reading of three views, or more, may well be necessary.

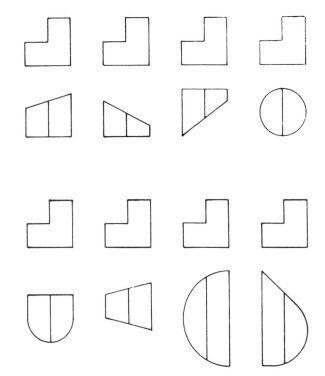

Fig. 4.16

Projection exercises

It is clear to us after teaching draughting and CAD for many years that visualizing a proposed new product in three dimensions, which is how you naturally view a finished article, is difficult when it is necessary to read more than one complex two dimensional drawing simultaneously. The draughtsman also ultimately needs to produce technically correct drawings, often from vague initial ideas. The very action of making proposal drawings stimulates many questions and their answers allow development to continue. Modifications to original ideas involve drawing amendments and changes to one view invariably have a 'knock on effect'. Comprehension, understanding and the ability to read technical drawings fluently comes with practice.

The following simple exercises are designed to assist in the perfection of draughting skills. They are equally suitable for CAD and the drawing board. Produce answers for each series and select standard sizes of drawing sheets, taking particular care with linework and layout.

If the CAD software program permits, move the separate views for each exercise so that they are positioned a similar distance from each other. Then experiment and position the groups to give a pleasing layout on the drawing sheet. Note how uniformity can improve presentation and give a professional appearance. Layout is a very important aspect when preparing drawings for desk top publishing applications.

1 Straight line examples

Figure 4.17 shows three components which have each been machined from solid blocks. These examples have been prepared on a grid formed by equilateral triangles. In every case, the scale is such that each side of the triangle will be 10 mm. For each component, draw five

(a)

(b)

(c)

Fig. 4.17

views in first angle projection, omitting hidden detail, and assume that the view in the direction of the arrow A will be the front view.

2   Examples involving radii and holes (Fig. 4.18)

For each example, project five views in first angle projection, taking the view in the direction of the arrow A as the front view. Hidden detail is required in the solutions to these problems, and note that in some cases the position of some of the holes will be found to coincide with centre lines. Where this occurs the dotted line should take priority. Take each side of the grid triangle to be 10 mm in length.

If only three views of each component were required, which ones would you choose? The professional draughtsman would select a front view, end view and plan view with the least number of dotted lines. Study your solutions carefully and where an ideal choice exists, box this solution with a thin chain line. In some cases more than one choice can be made and these are indicated in the solutions.

3   Examples with missing lines (first angle projection) (Fig. 4.19)

In the following projection examples, three views are given. Some views are incomplete with full lines and all dotted lines missing. Draw the given examples, using the scale provided. Complete each view, by inserting full lines where necessary and add all dotted lines to represent the hidden detail.

4   Examples with missing views (first angle projection) (Fig. 4.20)

In each of the following projection examples, two out of three views of simple solid components are shown. Draw the two views which are given using the scale provided. Complete each problem by drawing the missing view or plan in the space indicated by the cross.

5   First angle projection examples with plotted curves (Fig. 4.22)

In orthographic projection, all widths in the end view are equal in size to depths in the plan view, and of course the opposite is true that some dimensions required to complete end views may be obtained from given plan views. Fig. 4.21 shows part of a solid circular bar which has been cut at an angle of 30° with the horizontal axis. Point A is at any position along the sloping face. If a horizontal line is drawn through A across to the end view then the width of the chord is dimension X. This dimension is the distance across the cut face in the plan view and this has been marked on

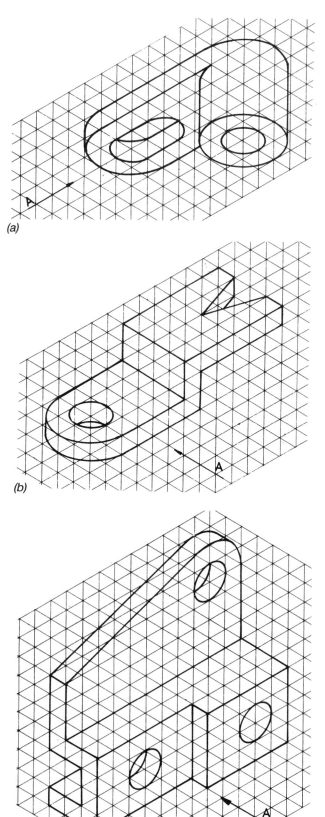

(a)

(b)

(c)

Fig. 4.18

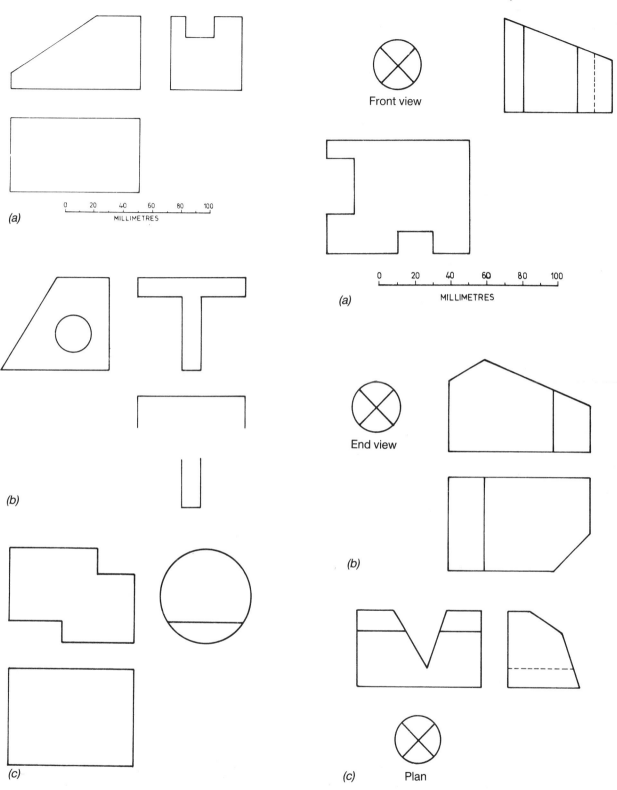

Front view

End view

Plan

(a)

(b)

(c)

MILLIMETRES

0 20 40 60 80 100

Fig. 4.19

Fig. 4.20

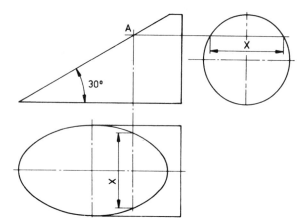

Fig. 4.21

the vertical line from A to the plan. If this procedure is repeated for other points along the sloping face in the front view then the resulting ellipse in the plan view will be obtained. All of the examples in this group may be solved by this simple method.

A word of warning: do not draw dozens of lines from points along the sloping face across to the end view and also down to the plan view before marking any dimensions on your solution. Firstly, you may be drawing more lines than you need, and in an examination this is a waste of time. Secondly confusion may arise if you accidently plot a depth on the wrong line. The golden rule is to draw one line, plot the required depth and then ask yourself 'Where do I now need other points to obtain an accurate curve?' Obviously, one needs to know in the plan view the position at the top and bottom of the slope, and the width at the horizontal centre line and at several points in between.

In the examples shown in Fig. 4.22 three views are given but one of them is incomplete and a plotted curve is required. Redraw each component using the scale provided. Commence each solution by establishing the extreme limits of the curve and then add intermediate points.

(a)

Fig. 4.22

0   20   40   60   80   100

MILLIMETRES

(b)

(c)

(d)

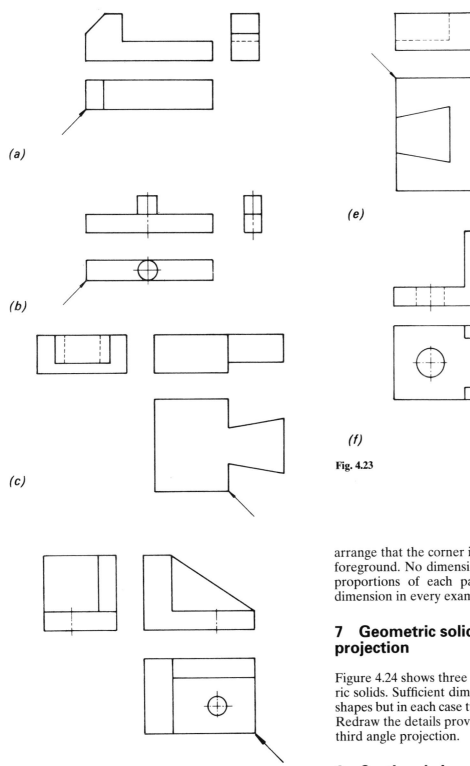

(a)

(b)

(c)

(d)

(e)

(f)

Fig. 4.23

6 Pictorial sketching from orthographic views

Figure 4.23 shows six components in first angle projection. Make a pictorial sketch of each component and arrange that the corner indicated by the arrow is in the foreground. No dimensions are given but estimate the proportions of each part assuming that the largest dimension in every example is 100 mm.

7 Geometric solids in third angle projection

Figure 4.24 shows three views of each of three geometric solids. Sufficient dimensions are given to define the shapes but in each case two of the views are incomplete. Redraw the details provided and complete the views in third angle projection.

8 Sectional views in third angle projection

In Fig. 4.25 there are three components and two views are provided for each one. Copy the views given, using the scale provided, and project the missing view which will be a section. Your solution should include the section plane, cross hatching and the statement A—A.

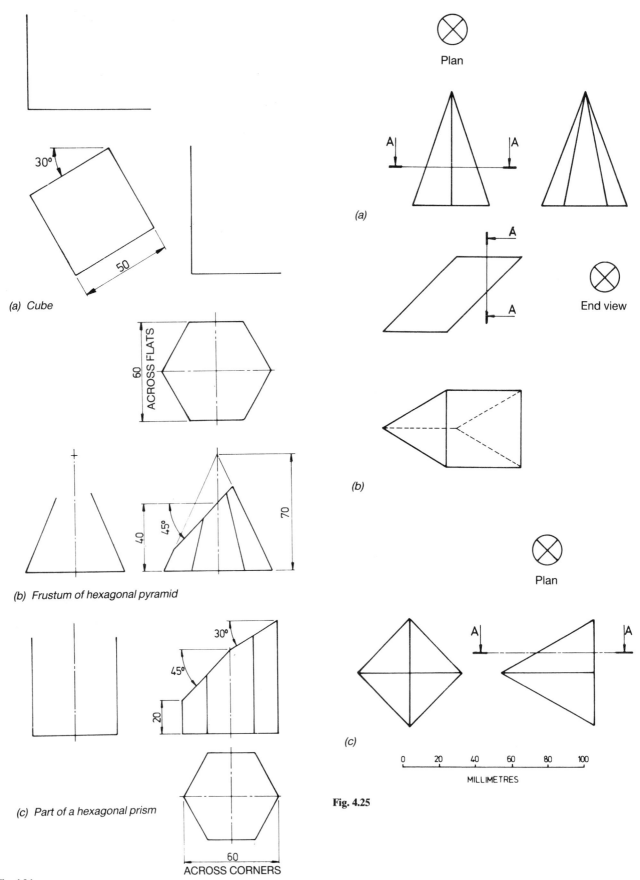

30°

50

(a) Cube

ACROSS FLATS
60

(b) Frustum of hexagonal pyramid

45°

70

40

45°

30°

20

(c) Part of a hexagonal prism

60
ACROSS CORNERS

Fig. 4.24

Plan

A |A

(a)

A

A

End view

(b)

Plan

A |A

(c)

0   20   40   60   80   100
MILLIMETRES

Fig. 4.25

(a)

(b)

(c)

(d)

Fig. 4.26

9 Dimensioning examples (first angle projection)

In Fig. 4.26 a scale is provided to enable each of the components to be redrawn. Redraw each example and add any dimensions which you consider necessary and which would be required by the craftsman. Bear in mind that if an object has sufficient dimensions to enable it to be drawn, then it can most likely be made. Hence, any sizes which are required to enable you to draw the part are also required by the manufacturer. For additional information regarding dimensioning refer to Chapter 14.

Chapter 5
Linework and lettering

Drawing paper sizes

The British Standard BS 308 has recommended that drawing sheets should be sized according to the ISO A series where for normal practical purposes the area of the largest sheet is one square metre and the sides are in the ratio of $1:\sqrt{2}$. The dimensions of the sheet are 841 mm $\times$ 1189 mm. For smaller sheets the longest side is progressively halved giving the designations and dimensions in Table 5.1. Since the A0 size has the area of one square metre, paper weights are conveniently expressed in the unit 'grams per square metre'.

Table 5.1

Designation	Size (millimetres)	Area
A0	841 × 1189	1 m^2
A1	594 × 841	5000 cm^2
A2	420 × 594	2500 cm^2
A3	297 × 420	1250 cm^2
A4	210 × 297	625 cm^2

Drawing sheets may be obtained from a standard roll of paper or already cut to size. Cut sheets sometimes have a border of at least 15 mm width to provide a frame and this frame may be printed with microfilm registration marks, which are triangular in shape and positioned on the border at the vertical and horizontal centre lines of the sheet.

Title blocks are also generally printed in the bottom right-hand corner of cut sheets and contain items of basic information required by the drawing office or user of the drawing. Typical references are as follows:

Name of firm,
Drawing number,
Component name,
Drawing scale and units of measurement,
Projection used (first or third angle) and or symbol,
Draughtsman's name and checker's signature,
Date of drawing and subsequent modifications,
Cross references with associated drawings or assemblies.

Other information will vary according to the branch and type of industry concerned but is often standardized by particular firms for their own specific purposes and convenience.

Lines and linework

Two thicknesses of line are recommended for manual and CAD drawings. A thick line and a thin line in the ratio of 2:1.

Standard lead holders, inking pens for manual use, and those for CAD plotters are all available in the following millimetre sizes: 0.25, 0.35, 0.5, 0.7, 1.0, 1.4 and 2.0.

Line thicknesses of 0.7 and 0.35 are generally used and will give good quality, black, dense and contrasting lines.

Table 5.2, shows applications for different line types which are designed to obtain a good professional finish to a drawing.

Various combinations of line thickness and type are shown on the mechanism in Fig. 5.2. (Circled numbers relate to the line types in Table 5.2.)

Fig. 5.3 shows part of a cone and if the complete cone was required, for example for dimensioning purposes, then the rest would be shown by adding thin continuous lines which intersect in a dot.

If it is necessary to show the initial outline of a part before it is bent or formed, then the initial outline can be indicated by a chain thin line which is double dashed. Fig. 5.4 shows the standard applied to a metal strip.

Fig. 5.5 shows the method of detailing a long strip of metal which has 60 holes in it at constant pitch. There would be no need to detail all of the component and this illustration gives one end only. The line to show the interruption in the drawing is thin continuous and with the zig-zag cutting line indicated by the letter J.

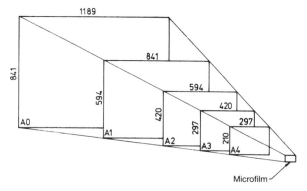

Fig. 5.1 Standard size reductions from A0 to 35 mm microfilm

Table 5.2 Types of line

Example	Application	
A	Continuous (thick)	Visible edges and outlines
B	Continuous (thin)	1 Dimension and projection lines 2 Hatching lines for cross sections 3 Leader lines 4 Outlines of revolved sections 5 Imaginary lines of intersection
C	Continuous (thin) irregular	Limits of partial views or sections provided the line is not an axis
D	Short dashes (thin)	Hidden outlines and edges
E	Chain (thin)	Centre lines
F	Chain (thick)	Used for surfaces which have to meet special requirements
G	Thin chain with thick line at ends and at changes in direction, to indicate cutting planes	Cutting planes which may be in one or more parallel planes, contiguous planes or intersecting planes. Refer to later chapter on 'Sections' for further examples
H	Chain thin with double dash	1 Preformed outlines 2 Adjacent parts 3 Extreme positions of moveable parts
J	Continuous straight thin line with zig zags	Limits of partial views; suitable for CAD drawings provided the line is not an axis

Fig. 5.2

Fig. 5.3   Example showing imaginary lines of intersection

Fig. 5.4   Initial outline application

Chain lines

Particular care should be taken with chain lines to ensure that they are neatly applied and attention is drawn to the following points:

a) All chain lines should start and finish with a long dash.
b) When centre points are defined, then the chain lines should cross one another at solid portions of the line.

c) Centre lines should extend for a short distance beyond the feature unless they are required for dimensioning or other purpose.
d) Centre lines should not extend through the spaces between views and should never terminate at another line on the drawing.

e) If an angle is formed by chain lines, then the long dashes should intersect and define the angle.

f) Arcs should meet straight lines at tangency points.

Fig. 5.5   Interrupted view application

g) When drawing hidden detail, a dashed line should start and finish with dashes in contact with the visible lines from which they originate.

h) Dashed lines should also meet with dashes at corners when drawing hidden detail.

Lettering

It has previously been mentioned that technical drawings are prepared using only two line thicknesses and if reasonable care is taken a pleasing result can easily be obtained. Drawings invariably need dimensions and notes and if these are added in a careless and haphazard manner, then a very poor overall impression may be given. Remember that technical drawings are the main line of communication between the originator and the user. Between a consultant and his client, the sales manager and his customer, the designer and the machine operator, a neat well executed technical drawing helps to establish confidence. The professional draughtsman also takes considerable pride in his work and much effort and thought is needed with respect to lettering, and spacing, in order to produce an acceptable drawing of high standard.

The following notes draw attention to small matters

of detail which we hope will assist the draughtsmen's technique of lettering.

1 Lettering may be vertical or slanted, according to the style which is customarily used by the draughtsman. The aim is to produce clear and unambiguous letters, numbers and symbols.
2 If slanted lettering is used, the slope should be approximately 65°–70° from the horizontal. Legibility is important. The characters should be capable of being produced at reasonable speed and in a repeatable manner. Different styles on the same drawing spoil the overall effect.
3 Use single stroke characters devoid of serifs and embellishments.
4 All strokes should be of consistent density.
5 The spacing round each character is important to ensure that 'filling in' will not occur during reproduction.
6 Lettering should not be underlined since this impairs legibility.
7 On parts lists or where information is tabulated, the letters or numerals should not be allowed to touch the spacing lines.
8 All drawing notes and dimensions should remain legible on reduced size copies and on the screens of microfilm viewers.
9 Capital letters are preferred to lower case letters since they are easier to read on reduced size copies of drawings. Lower case letters are generally used only where they are parts of standard symbols, codes or abbreviations.
10 When producing a manual drawing the draughtsman should take care to select the proper grade of pencil for lettering. The pencil should be sharp, but with a round point which will not injure the surface. Mechanical pencils save time and give consistent results since no resharpening is necessary.
11 Typewritten, stencilled or letters using the 'Letraset' adhesive letter system may be used since these provide uniformity and a high degree of legibility.

Minimum character height for capital letters and numerals

Table 5.3 gives the minimum recommended character heights for different sizes of drawing sheet and it is stressed that these are *minimum* sizes. If lower case letters are used then they should be proportioned so that the body height will be approximately 0.6 times the height of a capital letter.

The stroke thickness should be drawn at approximately 0.1 times the character height and the clear space between characters should be about 0.7 mm for 2.5 mm capitals and other sizes in proportion.

The spaces between lines of lettering should be consistent and preferably not less than half the character height. In the case of titles, this spacing may have to be reduced.

Table 5.3

Application	Drawing sheet size	Minimum character height
Drawing numbers etc.	A0, A1, A2 and A3	7 mm
	A4	5 mm
Dimensions and notes	A0	3.5 mm
	A1, A2, A3, and A4	2.5 mm

All notes should be placed so that they may be read from the same direction as the format of the drawing but there are cases, for example when a long vertical object is presented, where it may be necessary to turn the drawing sheet through 90° in the clockwise direction, in effect, to position the note which is then read from the right hand side of the drawing sheet.

The shape and form of an acceptable range of letters and numbers is illustrated in Fig. 5.6

Open styles are often used on drawings which are to be microfilmed, as increased clarity is obtainable on small reproductions.

(a) ABCDEFGHIJKLMNOPQRSTUVWXYZ
abcdefghijklmnopqrstuvwxyz
1234567890

(b) ABCDEFGHIJKLMNOPQRSTUVWXYZ
abcdefghijklmnopqrstuvwxyz
1234567890

(c) ABCDEFGHIJKLMNOPQRSTUVWXYZ
aabcdefghijklmnopqrstuvwxyz
1234567890

(d) ABCDEFGHIJKLMNOPQRSTUVWXYZ
aabcdefghijklmnopqrstuvwxyz
1234567890

(e) 1234567890

(f) 1234567890

Fig. 5.6

Erasing on manually produced drawings

It must be appreciated that after work has been undertaken on a drawing for a reasonable amount of time, then that drawing will possess some financial value. The draughtsman responsible for the drawing must be concerned with the reproducible quality of his work as prints or photographic copies are always taken from the originals. Erasings are not always undertaken due to drawing errors. Revisions and modifications are regularly made to update a product, due for example, to changes in materials, individual components, manufacturing techniques, operating experience and other causes outside the draughtsman's control. The point to remember is that on the area covered by the erasure, part of a new drawing will be added and the quality of this drawing must be identical in standard with the original. Obviously, if the surface of the drawing sheet has been damaged in any way during erasure, then the draughtsman performing the work must start with a serious disadvantage.

The following suggestions are offered to assist in the preservation of drawings when erasures have to be made.

1   Choose erasers with much care. Line removal without damaging the drawing surface is essential.
2   An erasing shield will assist in the protection of the areas adjacent to the modifications and eliminate smudging other lines which are not being removed.
3   Do not use highly abrasive erasers, some tend to wrinkle the surface due to the effects of heat from excessive friction.
4   Avoid the use of razor blades and knives.
5   Thoroughly erase the lines, as a ghost effect can be observed with incomplete erasures when the prints are made.
6   It sometimes helps to place a hard surface under the area being erased. A flat plastic set square under the drawing sheet will assist in raising the fibres of the paper. When drawings are originally made, the action of the pencil is bound to produce grooves in the drawing sheet and these can often be ironed out, if erasures are undertaken on a flat hard surface.
7   After erasing, clean the back of the sheet as pressure on the front side of the drawing often transfers dirt containing a high proportion of graphite on to the back of the sheet. If complete erasure proves difficult, check that the offending marks are definitely on the front of the sheet. Graphite marks may well have been made on the back when the original was produced by pressure through the drawing on to a dirty backing sheet on the draughting machine.
8   A draughtsman uses a brush to remove the eraser crumbs from the surface of the drawing and this prevents smudging.

If in any doubt, a little time spent performing experimental trial erasures on a scrap of a similar drawing medium will pay dividends–far better than experimenting on a valuable original.

The care and storage of original drawings

Drawings are valuable and it is very important that adequate and satisfactory handling and storage facilities should be made available in order to preserve them in first class condition. Drawings may be used and re-used many times and minimum wear and tear is essential if good reproductions and microfilms are to be obtained over a long period of time. The following simple rules will assist in keeping drawings in 'mint' condition.

1   Never fold drawings.
2   Don't use your hands for wiping off dirt and debris; use a draughtsman's brush.
3   Dirt also collects on the reverse side of drawings so keep them as clean as the top side.
4   Apart from the period when the drawing is being prepared or modified, it is good policy to refer to prints at other times when the drawing is required for information purposes.
5   The drawing board should be covered at all times outside normal office hours, to avoid the collection of dust and dirt.
6   Too many drawings should not be crowded in a filing drawer. Most drawing surfaces, paper or plastics, are reasonably heavy and damage results from careless manipulation in and out of drawers.
7   Do not roll drawings tightly since they may not lie flat during microfilming.
8   Do not use staples or drawing pins. Tape and drawing clips are freely available.
9   Try to keep all draughting equipment clean, and lift, rather than slide them, across the drawing surface.
10 When using drawings, try and use a large reference table. Lift the drawings rather than slide them, to avoid smudging and wear.
11 Advice regarding erasures is given in the section above and should be followed.
12 Perspiration can easily be transferred to a drawing if it is leaned on during very humid weather and this increases the pickup of dirt.
13 When filing drawings, it is recommended that similar sizes are filed together. The files should if possible be fireproof and if a fire sprinkler system is installed in the drawing office or filing area then it should be a regular practice to keep the cabinets closed at night and at weekends when the office is not occupied.
14 Drawings should be stored under conditions of normal heat and humidity, about 21°C and 40 to 60% relative humidity.

Chapter 6

Three dimensional illustrations using isometric and oblique projection

Isometric projection

Fig. 6.1 shows three views of a cube in orthographic projection; the phantom line indicates the original position of the cube, and the full line indicates the position after rotation about the diagonal AB. The cube has been rotated so that the angle of 45° between side AC_1 and diagonal AB now appears to be 30° in the front elevation, C_1 having been rotated to position C. It can clearly be seen in the end view that to obtain this result the angle of rotation is greater than 30°. Also, note that, although DF in the front elevation appears to be vertical, a cross check with the end elevation will

confirm that the line slopes, and that point F lies to the rear of point D. However, the front elevation now shows a three dimensional view, and when taken in isolation it is known as an *isometric projection*.

This type of view is commonly used in pictorial presentations, for example in car and motor-cycle service manuals and model kits, where an assembly has been 'exploded' to indicate the correct order and position of the component parts.

It will be noted that, in the isometric cube, line AC_1 is drawn as line AC, and the length of the line is reduced. Fig. 6.2 shows an isometric scale which in principle is obtained from lines at 45° and 30° to a horizontal axis.

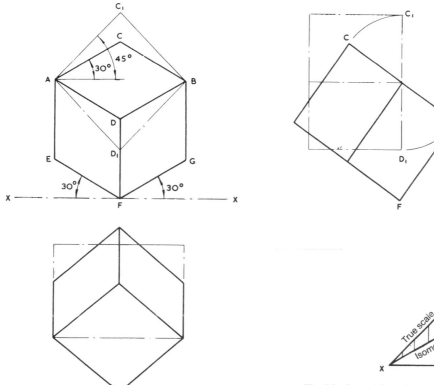

Fig. 6.1

Fig. 6.2 Isometric scale

The 45° line XY is calibrated in millimetres commencing from point X, and the dimensions are projected vertically on to the line XZ. By similar triangles, all dimensions are reduced by the same amount, and isometric lengths can be measured from point X when required. The reduction in length is in the ratio

$$\frac{isometric\ length}{true\ length} = \frac{\cos 45°}{\cos 30°} = \frac{0.7071}{0.8660} = 0.8165$$

Now, to reduce the length of each line by the use of an isometric scale is an interesting academic exercise, but commercially an isometric projection would be drawn using the true dimensions and would then be enlarged or reduced to the size required.

Note that, in the isometric projection, lines AE and DB are equal in length to line AD; hence an equal reduction in length takes place along the apparent vertical and the two axes at 30° to the horizontal. Note also that the length of the diagonal AB does not change from orthographic to isometric, but that of diagonal C_1D_1 clearly does. When setting out an isometric projection, therefore, measurements must be made only along the isometric axes EF, DF, and GF.

Fig. 6.3 shows a wedge which has been produced from a solid cylinder, and dimensions A, B, and C indicate typical measurements to be taken along the principle axes when setting out the isometric projection. Any curve can be produced by plotting a succession of points in space after taking ordinates from the X, Y, and Z axes.

Fig. 6.4 (a) shows a cross-section through an extruded alloy bar: the views (b), (c), and (d) give alternative isometric presentations drawn in the three principal planes of projection. In every case, the lengths of ordinates OP, OQ, P1, and Q2, etc. are the same, but are positioned either vertically or inclined at 30° to the horizontal.

Fig. 6.5 shows an approximate method for the construction of isometric circles in each of the three major planes. Note the position of the points of intersection of radii RA and RB.

The construction shown in Fig. 6.5 can be used partly for producing corner radii. Fig. 6.6 shows a small block with radiused corners together with isometric projection which emphasises the construction to find the centres for the corner radii; this should be the first part of the drawing to be attempted. The thickness of the block is obtained from projecting back these radii a distance equal to the block thickness and at 30°. Line in those parts of the corners visible behind the front face, and complete the pictorial view by adding the connecting straight lines for the outside of the profile.

In the approximate construction shown, a small inaccuracy occurs along the major axis of the ellipse, and fig. 6.7 shows the extent of the error in conjunction with a plotted circle. In the vast majority of applications where complete but small circles are used, for example spindles, pins, parts of nuts, bolts, and fixing holes, this error is of little importance and can be neglected.

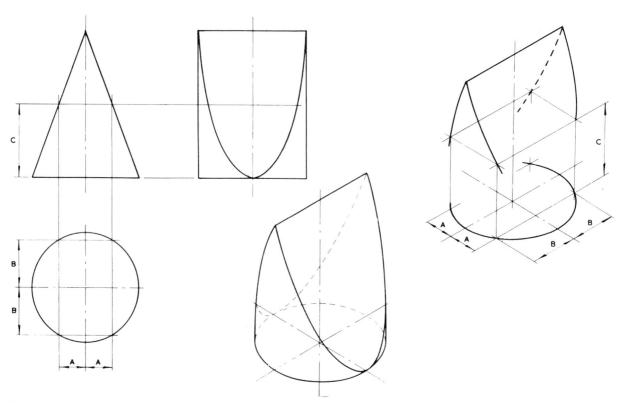

Fig. 6.3 Construction principles for points in space, with complete solution

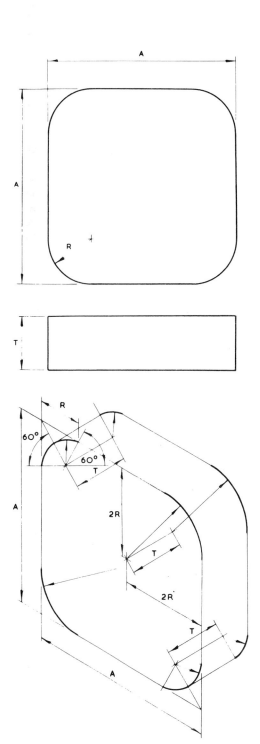

Fig. 6.4 Views (b), (c) and (d) are isometric projections of the section in view (a)

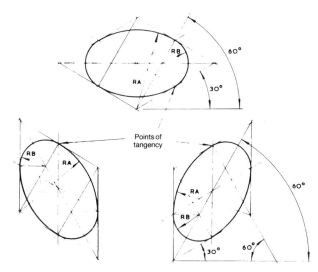

Fig. 6.5

Fig. 6.6   Isometric constructions for corner radii

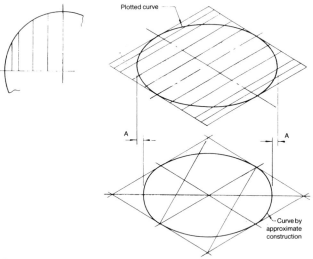

Fig. 6.7 Relationship between plotted points and constructed isometric circles

Oblique projection

Fig. 6.8 shows part of a plain bearing in orthographic projection, and Figs 6.9 and 6.10 show alternative pictorial projections.

It will be noted in Figs 6.9 and 6.10 that the thickness of the bearing has been shown by projecting lines at 45° back from a front elevation of the bearing. Now, Fig. 6.10 conveys the impression that the bearing is thicker than the true plan suggests, and therefore in Fig. 6.9 the thickness has been reduced to one half of the actual size. Fig. 6.9 is known as an *oblique projection*, and objects which have curves in them are easiest to draw if they are turned, if possible, so that the curves are presented in the front elevation. If this proves impossible or undesirable, then Fig. 6.11 shows part of

the ellipse which results from projecting half sizes back along the lines inclined at 45°.

A small die-cast lever is shown in Fig. 6.12, to illustrate the use of a reference plane. Since the bosses are of different thicknesses, a reference plane has been taken along the side of the web; and, from this reference plane, measurements are taken forward to the boss faces and backwards to the opposite sides. Note that the points of tangency are marked, to position the slope of the web accurately.

With oblique and isometric projections, no allowance is made for perspective, and this tends to give a slightly unrealistic result, since parallel lines moving back from the plane of the paper do not converge.

Fig. 6.11

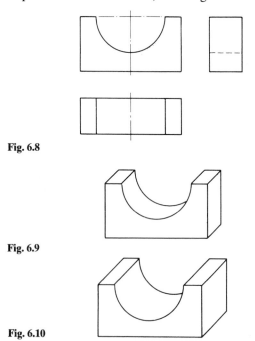

Fig. 6.8

Fig. 6.9

Fig. 6.10

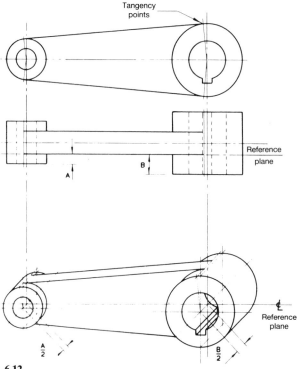

Fig. 6.12

Chapter 7
Drawing layouts and simplified methods

Single-part drawing

A single-part drawing should supply the complete detailed information to enable a component to be manufactured without reference to other sources. It should completely define shape or form and size, and should contain a specification. The number of views required depends on the degree of complexity of the component. The drawing must be fully dimensioned, including tolerances where necessary, to show all sizes and locations of the various features. The specification for the part includes information relating to the material used and possible heat-treatment required, and notes regarding finish. The finish may apply to particular surfaces only, and may be obtained by using special machining operations or, for example, by plating, painting, or enamelling. Fig. 7.1 shows typical single-part drawings.

An alternative to a single-part drawing is to collect several small details relating to the same assembly and group them together on the same drawing sheet. In practice, grouping in this manner may be satisfactory provided all the parts are made in the same department, but it can be inconvenient where, for example, pressed parts are drawn with turned components or sheet-metal fabrications.

More than one drawing may also be made for the same component. Consider a sand-cast bracket. Before the bracket is machined, it needs to be cast; and, before casting, a pattern needs to be produced by a pattern-maker. It may therefore be desirable to produce a drawing for the patternmaker which includes the various machining allowances, and then produce a separate drawing for the benefit of the machinist which shows only dimensions relating to the surfaces to be machined and the size of the finished part. The two drawings would each have only parts of the specification which suited one particular manufacturing process. (See also Figs 14.34 and 14.35.)

Collective single-part drawings

Fig. 7.2 shows a typical collective single-part drawing for a rivet. The drawing covers 20 rivets similar in every respect except length; in the example given, the part number for a 30 mm rivet is S123/13. This type of drawing can also be used where, for example, one or two dimensions on a component (which are referred to on the drawing as A and B) are variable, all others being standard. For a particular application, the draughtsman would insert the appropriate value of dimensions A and B in a table, then add a new suffix to the part number. This type of drawing can generally be used for basically similar parts.

Assembly drawings

Machines and mechanisms consist of numerous parts, and a drawing which shows the complete product with all its components in their correct physical relationship is known as an assembly drawing. A drawing which gives a small part of the whole assembly is known as a sub-assembly drawing. A sub-assembly may in fact be a complete unit it self; for example, a drawing of a clutch could be considered as a sub-assembly of a drawing showing a complete automobile engine. The amount of information given on an assembly drawing will vary considerably with the product and its size and complexity.

If the assembly is relatively small, information which might be given includes a parts list. The parts list, as the name suggests, lists the components, which are numbered. Numbers in 'balloons' with leader lines indicate the position of the component on the drawing (see Fig. 7.3). The parts list will also contain information regarding the quantity required of each component for the assembly, its individual single-part drawing number, and possibly its material. Parts lists are not standard items, and their contents vary from one drawing office to another.

The assembly drawing may also give other information, including overall dimensions of size, details of bolt sizes and centres where fixings are necessary, weights

Fig. 7.1(a)   Bearing insert

Fig. 7.1(b)   Gear hub

Fig. 7.1(c)   Retaining ring

Fig. 7.2   Collective single-part drawing of a rivet

Fig. 7.3 Assembly drawing of gear and bearing

ITEM No	TITLE	No. OFF	PART No.
1	BEARING INSERT	1	0001
2	GEAR HUB	1	0002
3	RETAINING RING	1	0003

COPYRIGHT NOTE	ASSEMBLY OF GEAR & BEARING	ORIGINAL SCALE 1:1
	MATERIAL ——	PART No 0004

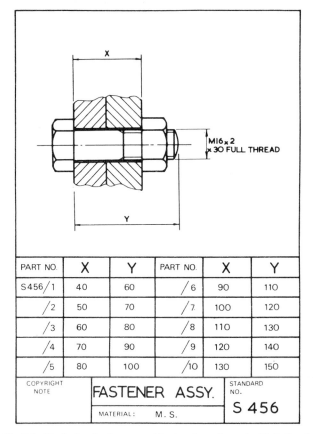

Fig. 7.4 Typical collective assembly drawing of a nut with bolts of various lengths

PART NO.	X	Y	PART NO.	X	Y
S456/1	40	60	/6	90	110
/2	50	70	/7	100	120
/3	60	80	/8	110	130
/4	70	90	/9	120	140
/5	80	100	/10	130	150

COPYRIGHT NOTE	FASTENER ASSY.	STANDARD NO.
	MATERIAL: M. S.	S 456

required for shipping purposes, operating details and instructions, and also, perhaps, some data regarding the design characteristics.

Collective assembly drawing

This type of drawing is used where a range of products which are similar in appearance but differing in size is manufactured and assembled. Fig. 7.4 shows a nut-and-bolt fastening used to secure plates of different combined thickness; the nut is standard, but the bolts are of different lengths. The accompanying table is used to relate the various assemblies with different part numbers.

Design layout drawings

Most original designs are planned in the drawing office where existing or known information is collected and used to prepare a provisional layout drawing before further detailed design work can proceed. This type of drawing is of a preliminary nature and subject to much modification so that the designer can collect his

thoughts together. The drawing can be true to scale or possibly enlargements or reductions in scale depending on the size of the finished product or scheme, and is essentially a planning exercise. They are useful in order to discuss proposals with prospective customers or design teams at a time when the final product is by no means certain, and should be regarded as part of the design process.

Provisional layout drawings may also be prepared for use with tenders for proposed work where the detailed design will be performed at a later date when a contract has been negotiated, the company being confident that it can ultimately design and manufacture the end product. This confidence will be due to experience gained in similar schemes undertaken previously.

Combined detail and assembly drawings

It is sometimes convenient to illustrate details with their assembly drawing on the same sheet. This practice is particularly suited to small 'one-off' or limited-production-run assemblies. It not only reduces the actual number of drawings, but also the drawing-office time spent in scheduling and printing. Fig. 7.5 shows a simple application of an assembly of this type.

Fig. 7.5   Combined detail and assembly drawing of hub-puller

Exploded assembly drawings

Fig. 7.6 shows a typical exploded assembly drawing; these drawings are prepared to assist in the correct understanding of the various component positions in an assembly. Generally a pictorial type of projection is used, so that each part will be shown in three dimensions. Exploded views are invaluable when undertaking servicing and maintenance work on all forms of plant and appliances. Car manuals and do-it-yourself assembly kits use such drawings, and these are easily understood. As well as an aid to construction, an exploded assembly drawing suitably numbered can also be of assistance in the ordering of spare parts; components are more easily recognisable in a pictorial projection, especially by people without training in the reading of technical drawings.

Simplified drawings

Simplified draughting conventions have been devised to reduce the time spent drawing and detailing symmetrical components and repeated parts. Fig. 7.7 shows a gasket which is symmetrical about the horizontal centre line. A detail drawing indicating the line of symmetry and half of the gasket is shown in Fig. 7.8, and this is sufficiently clear for the part to be manufactured.

If both halves are similar except for a small detail, then the half which contains the exception is shown with an explanatory note to that effect, and a typical example is illustrated in Fig. 7.9.

A joint-ring is shown in Fig. 7.10, which is symmetrical about two axes of symmetry. Both axes are shown in the detail, and a quarter view of the joint-ring is sufficient for the part to be made.

The practice referred to above is not restricted to flat thin components, and Fig. 7.11 gives a typical detail of a straight lever with a central pivot in part section. Half the lever is shown, since the component is symmetrical, and a partial view is added and drawn to an enlarged scale to clarify the shape of the boss and leave adequate space for dimensioning.

Repeated information also need not be drawn in full; for example, to detail the peg-board in Fig. 7.12 all that is required is to draw one hole, quoting its size and fixing the centres of all the others.

Similarly, Fig. 7.13 shows a gauze filter. Rather than draw the gauze over the complete surface area, only a small portion is sufficient to indicate the type of pattern required.

Knurled screws are shown in Fig. 7.14 to illustrate the accepted conventions for straight and diamond knurling.

Fig. 7.6

Fig. 7.7   **Fig. 7.8**

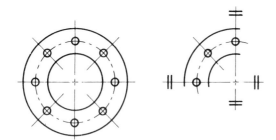

Fig. 7.9 When dimensioning add drawing note 'slot on one side only'

Fig. 7.10

Fig. 7.12

Fig. 7.13

Example of   Example of
straight knurling   diamond knurling

Fig. 7.14

SCALE 2:1

PART SECTION XX

Fig. 7.11 Part of a lever detail drawing symmetrical about the horizontal axis

Machine drawing

The draughtsman must be able to appreciate the significance of every line on a machine drawing. He must also understand the basic terminology and vocabulary used in conjunction with machine drawings.

Machine drawings of components can involve any of the geometrical principles and constructions described in this book and in addition the accepted drawing standards covered by BS 308.

Fig. 7.15 illustrates many features found on machine drawings and the notes which follow give additional explanations and revision comments.

1 *Angular dimension*—Note that the circular dimension line is taken from the intersection of the centre lines of the features.
2 *Arrowheads*—The point of an arrowhead should touch the projection line or surface, it should be neat and easily readable and normally not less than 3 mm in length.
3 *Auxiliary dimension*—A dimension given for information purposes but not used in the actual manufacturing process.
4 *Boss*—A projection, which is usually circular in cross section, and often found on castings and forgings. A shaft boss can provide extra bearing support, for example, or a boss could be used on a thin cast surface to increase its thickness in order to accommodate a screw thread.
5 *Centre line*—A thin chain line which is used to indicate the axes of holes, components and circular parts.
6 *Thick chain line*—This is used to indicate surfaces which are required to meet special specifications and which differ from the remainder of the component.
7 *Chamfer*—A chamfer is machined to remove a sharp edge. The angle is generally 45°. Often referred to as a bevelled edge.
8 *Circlip groove*—A groove to accommodate a circlip. A circlip may be manufactured from spring steel wire, sheet or plate which is hardened and tempered and when applied in an assembly provides an

Fig. 7.15

inward or outward force to locate a component within a bore or housing.

9 *Clearance hole*—A term used in an assembly to describe a particular hole which is just a little larger and will clear the bolt or stud which passes through.

10 *Counterbore*—A counterbored hole may be used to house a nut or bolthead so that it does not project above a surface. It is machined so that the bottom surface of the larger hole is square to the hole axis.

11 *Countersink*—A hole which is recessed conically to accommodate the head of a rivet or screw so that the head will lie at the same level as the surrounding surface.

12 *Section plane* or *cutting plane*—These are alternative terms used to define the positions of planes from which sectional elevations and plans are projected.

13 *Dimension line*—This is a thin continuous line which is placed outside the outline of the object, if possible. The arrowheads touch the projection lines. The dimension does not touch the line but is placed centrally above it.

14 *Enlarged view*—Where detail is very small or insufficient space exists for dimensions or notes then a partial view may be drawn with an increased size scale.

15 *Round*—This term is often used to describe an external radius.

16 *Fillet*—This is the term given to the radii on internal corners. Often found on castings, where its function is to prevent the formation of stress cracks, which can originate from sharp corners. Where three surfaces meet on a casting the fillet radii will be spherical.

17 *Flange*—This is a term to describe a projecting rim or an edge which is used for stiffening or for fixing.

The example here is drilled for countersunk screws.

18 *Hatching*—Note that cross hatching of the component at the section plane is performed with thin continuous lines at 45°. Spacing between the hatching lines varies with the size of the component but should not be less than 4 mm.

19 *Hidden detail*—Indicated by thin chain line. The line should comprise dashes of consistent length and spacing. Dashes of 3 mm and spaces of 2 mm are of reasonable proportion.

20 *Knurl*—A surface finish with a square or diamond pattern. Can be used in a decorative manner or to improve grip.

21 *Leader line*—Leaders are used to indicate where dimensions or notes apply and are drawn as thin continuous lines terminating in arrowheads or dots. An arrowhead should always terminate on a line; dots should be within the outline of the object.

22 *Local section*—A local section may be drawn if a complete section or a half section is inconvenient. The local break line around the section is a continuous thin irregular line.

23 *Machining centre*—An accurately drilled hole with a good finish at each end of the component which enables the work to be located during a machining operation on a lathe.

24 *Machining symbol*—If it is desired to indicate that a particular surface is to be machined, without further defining the actual machining process or the surface finish, a symbol is added normal to the line representing the surface. The included angle of the symbol is approximately 60°. A general note may be added to a drawing where all surfaces are to be machined as follows:

25 *Surface finish*—If a surface is to be machined and a particular quality surface texture is desired then a standard machining symbol is added to the drawing with a number which gives the maximum permissible roughness expressed numerically in micrometres.

26 *Surface finish*—If maximum and minimum degrees of roughness are required then both figures are added to the machining symbol.

27 *Pitch circle diameter*—A circle which passes through the centres of a series of holes. The circle is drawn as a thin chain line.

28 *Recess*—A hollow feature which is used to reduce the overall weight of the component. A recess can also be used to receive a mating part.

29 *Slot*—An alternative term to a slit, groove, channel or aperture.

30 *Spigot*—This is a circular projection which is machined to provide an accurate location between assembled components.

31 *Splined shaft*—A rotating member which can transmit a torque to a mating component. The mating component may move axially along the splines which are similar in appearance to keyways around the spindle surface.

32 *Square*—Diagonal lines are drawn to indicate the flat surface of the square and differentiate between a circular and a square section shaft. The same convention is used to show spanner flats on a shaft.

33 *Taper*—A term used in connection with a slope or incline. Rate of taper can also define a conical form.

34 *Taper symbol*—The taper symbol is shown here in a rectangular box which also includes dimensional information regarding the rate of taper on the diameter.

35 *External thread*—An alternative term used for a male thread. The illustration here shows the thread convention.

36 *Internal thread*—An alternative term for a female thread. The illustration here shows the convention for a female tapped hole.

37 *Undercut*—A circular groove at the bottom of a thread which permits assembly without interference from a rounded corner. Note in the illustration that a member can be screwed along the M20 thread right up to the tapered portion.

38 *Woodruff key*—A key shaped from a circular disc which fits into a circular keyway in a tapered shaft. The key can turn in the circular recess to accommodate any taper in the mating hub.

39 *Key*—A small block of metal, square or rectangular in cross section, which fits between a shaft and a hub and prevents circumferential movement.

40 *Keyway*—A slot cut in a shaft or hub to accommodate a key.

Drawing scales

Small objects are sometimes drawn larger than actual size, while large components and assemblies, of necessity, are drawn to a reduced size. A drawing should always state the scale used. The scale on a full-size drawing will be quoted as 'ORIGINAL SCALE 1:1'.

Drawings themselves should not be scaled when in use for manufacturing purposes, and warnings against the practice are often quoted on standard drawing sheets, e.g. 'DO NOT SCALE' and 'IF IN DOUBT, ASK'. A drawing must be adequately dimensioned, or referenced sufficiently so that all sizes required are obtainable.

The recommended multipliers for scale drawings are 2, 5, and 10.

    1:1 denotes a drawing drawn full-size.
    2:1 denotes a drawing drawn twice full-size.
    5:1 denotes a drawing drawn five times full size.

Other common scales are 10:1, 20:1, 50:1, 100:1, 200:1, 500:1, and 1000:1.

It should be pointed out that a scale drawing can be deceiving; a component drawn twice full-size will cover four times the area of drawing paper as the same component drawn full-size, and its actual size may be difficult to visualize. To assist in appreciation, it is a common practice to add a pictorial view drawn full-size, provided that the drawing itself is intended to be reproduced to the same scale and not reproduced and reduced by microfilming.

The recommended divisors for scale drawings are also 2, 5, and 10.

> 1:1 denotes a drawing drawn full-size.
> 1:2 denotes a drawing drawn half full-size.
> 1:5 denotes a drawing drawn a fifth full-size.

Other common scales used are 1:10, 1:20, 1:50, 1:100, 1:200, 1:500, and 1:1000.

The draughtsman will select a suitable scale to use on a standard drawing sheet and this will depend on the size of the object to be drawn. Remember that the drawing must clearly show necessary information and detail. It may be beneficial to make a local enlargement of a small area and an example is given in Fig. 7.15.

Scale used in geometric construction

Division of lines

Fig. 7.16 shows the method of dividing a given line AB, 89 mm long, into a number of parts (say 7).

Draw line AC, and measure 7 equal divisions. Draw line B7, and with the tee-square and set-square draw lines parallel to line B7 through points 1 to 6, to give the required divisions on AB.

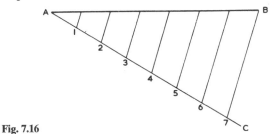

Fig. 7.16

Fig. 7.17 shows an alternative method.

1 Draw vertical lines from A and B.
2 Place the scale rule across the vertical lines so that seven equal divisions are obtained and marked.
3 Draw vertical lines up from points 2 to 7 to intersect AB.

Fig. 7.17

Fig. 7.18 Diagonal scale where 40 mm represents 1000 mm

Diagonal scales

Fig. 7.18 shows the method of drawing a diagonal scale of 40 mm to 1000 which can be read by 10 mm to 4000. Diagonal scales are so called since diagonals are drawn in the rectangular part at the left-hand end of the scale. The diagonals produce a series of similar triangles.

1 Draw a line 160 mm long.
2 Divide the line into four equal parts.
3 Draw 10 vertical divisions as shown and to any reasonable scale (say 5 mm) and add diagonals.

An example of reading the scale is given.

Fig. 7.19 Plain scale where 30 mm represents 500 mm

Plain scales

The method of drawing a plain scale is shown in fig. 7.19. The example is for a plain scale of 30 mm to 500 mm to read by 125 mm to 2500.

1 Draw a line 150 mm long and divide it into 5 equal parts.
2 Divide the first 30 mm length into four equal parts, and note the zero position on the solution.

An example of a typical reading is given.

This method of calibration is in common use in industry, and scales can be obtained suitable for a variety of scale ratios.

Abbreviations

In order to shorten drawing notes we often use abbreviations, and the following list gives a selection of terms in accordance with current British Standards.

Term	Abbreviation or symbol
Across flats	A/F
Assembly	ASSY
Centres	CRS
Centre line	℄ or CL
Chamfered	CHAM
Cheese head	CH HD
Countersunk	CSK
Countersunk head	CSK HD
Counterbore	C'BORE
Cylinder or cylindrical	CYL
Diameter (in a note)	DIA
Diameter (preceding a dimension)	∅
Drawing	DRG
External	EXT
Figure	FIG.
Hexagon	HEX
Hexagon head	HEX HD
Insulated or insulation	INSUL
Internal	INT
Left hand	LH
Long	LG
Material	MATL
Maximum	MAX
Minimum	MIN
Number	NO.
Pattern number	PATT NO.
Pitch circle diameter	PCD
Radius (preceding a dimension, capital letter only)	R
Required	REQD
Right hand	RH
Round head	RD HD
Screwed	SCR
Sheet	SH
Sketch	SK
Specification	SPEC
Spherical diameter (preceding a dimension)	S∅
Spherical radius (preceding a dimension (∅)	SR
Spotface	S'FACE
Square (in a note)	SQ
Square (preceding a dimension)	□ or ⊠
Standard	STD
Undercut	U'CUT
Volume	VOL
Weight	WT

Chapter 8
Sections and sectional views

A section is used to show the detail of a component, or an assembly, on a particular plane which is known as the cutting plane. A simple bracket is shown in Fig. 8.1 and it is required to draw three sectional views. Assume that you had a bracket and cut it with a hacksaw along the line marked B—B. If you looked in the direction of the arrows then the end view B—B in the solution (Fig. 8.2), would face the viewer and the surface indicated by the cross hatching would be the actual metal which the saw had cut through. Alternatively had we cut along the line C—C then the plan in the solution would be the result. A rather special case exists along the plane A—A where in fact the thin web at this point has been sliced. Now if we were to cross hatch all the surface we had cut through on this plane we would give a false impression of solidity. To provide a more realistic drawing the web is defined by a full line and the base and perpendicular parts only have been cross hatched. Note, that cross hatching is never undertaken between dotted lines, hence the full line between the web and the remainder of the detail. However, the boundary at this point is theoretically a dotted line since the casting is formed in one piece and no join exists here. This standard drawing convention is frequently tested on examination papers.

Cutting planes are indicated on the drawing by a long chain line 0.35 mm thick and thickened at both ends to 0.7 mm. The cutting plane is lettered and the arrows indicate the direction of viewing. The sectional view or plan must then be stated to be A—A, or other letters appropriate to the cutting plane. The cross hatching should always be at 45° to the centre lines, with continuous lines 0.35 mm thick.

If the original drawing is to be microfilmed successive lines should not be closer than 4 mm as hatching lines tend to merge with much reduced scales. When hatching very small areas the minimum distance between lines should not be less than 1 mm.

In the case of very large areas, cross hatching may be limited to a zone which follows the contour of the hatched area. On some component detail drawings it may be necessary to add dimensions to a sectional

Fig. 8.1

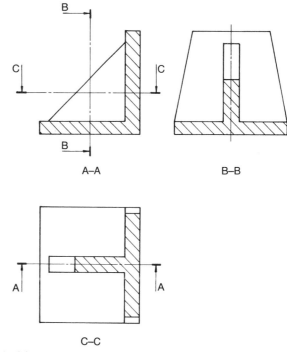

Fig. 8.2

drawing and the practice is to interrupt the cross hatching so that the letters and numbers are clearly visible.

Fig. 8.3 shows three typical cases of cross hatching. Note that the hatching lines are equally spaced and drawn at an angle of 45° to the principal centre line in each example.

Fig. 8.3

A bush is shown in Fig. 8.4 in a housing. There are two adjacent parts and each is cross hatched in opposite directions. It is customary to reduce the pitch between hatching lines for the smaller part.

Fig. 8.4

If the interior of a component is of an intricate nature or it contains several parts to form an assembly, then the customary orthographic drawing would contain a confusion of dotted lines, which, in addition to being difficult to draw could also be terribly difficult to understand. The reader of any engineering drawing should be able to obtain only one positive interpretation of the component, or the draughtsman has failed in his duty. Sectional drawings are prepared which cut away a portion of the component to reveal internal details and certain standard conventions have been established to cover this aspect of drawing practice.

Fig. 8.5 shows some advantages of drawing a sectional view with a small cast component.

Note, that in Plan (A), the sectional plan gives clearly the exact outline along the horizontal axis where the casting has assumed to have been cut. This contrasts with the confusion in Plan (B) which obviously results from attempting to include all the detail by inserting the appropriate dotted lines.

Where the location of a single cutting plane is obvious, no indication of its position or identification is required. Fig. 8.6 gives a typical example.

Half sections

Symmetrical parts may be drawn half in section and half in outside view. This type of drawing avoids the necessity of introducing dotted lines for the holes and the recess. Dimensioning to dotted lines is not a recommended practice.

Fig. 8.5

Plan A

A–A

Plan B

Fig. 8.6

Fig. 8.7

Revolved sections

A special spanner is illustrated in Fig. 8.8. A revolved section is shown on the handle to indicate the shape of the cross section at that point. This is a convenient convention to use on single view drawings because the

Fig. 8.8

Fig. 8.9

Fig. 8.11

shape could not be confirmed without projecting a second view or an added note.

A second type of revolved section in Fig. 8.9 shows a case where it is required to indicate details on two separate intersecting planes. The elevation in section has been drawn assuming that the right hand plane has been revolved to the horizontal position. Note that the thin web is not cross hatched.

Fig. 8.10

Fig. 8.10 shows a sectioned elevation from a plan where the section line is taken along three neighbouring planes which are not at right angles to one another. The section line follows the section planes in order, and is thickened at each change of direction.

Removed sections

A removed section is shown in Fig. 8.11. Note that no additional background information has been included, since the removed section only indicates the true shape of the casting at the point where the section has been taken. B—B gives the section along the horizontal centre line through the thin web.

Sections through thin material

Many products are manufactured from very thin materials which would be virtually impossible to cross hatch in a sectional view and in these cases it is usual to make them entirely black. Where however two or more thin sections are adjacent to each other, a gap is left so that the profile of the separate components is clearly defined. A compound stanchion used in structural steelwork and drawn to reduced scale is shown in Fig. 8.12. The same situation applies with sections through sheet-metal fabrications, gaskets, seals and packings.

Fig. 8.12

Local sections

It is not always necessary to draw a complete section through a component if a small amount of detail only needs to be illustrated. A typical example is shown in Fig. 8.13 where a keyway is drawn in a section. The irregular line defines the boundary of the section. It is not required to add a section plane to this type of view.

Fig. 8.13

Components not drawn in section

It is the custom not to section many recognizable components in assembly drawings positioned along the cutting plane; these include nuts, bolts, washers, rivets, pins keys, balls, rollers, spokes of wheels and similar symmetrical parts.

Successive sections

Fig. 8.14 shows the front and end elevations of a special purpose mounting plate where sectional plans are given at different levels to illustrate the shapes of the various cutouts and details. Now it will be noted that the presentation of this problem takes considerable vertical space since all of the plan views are in correct projection. The current British Standard permits successive sections to be drawn as shown in Fig. 8.14.

Fig. 8.14

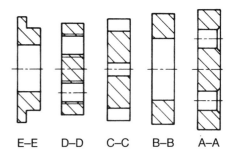

E–E   D–D   C–C   B–B   A–A

Fig. 8.15

Note that where successive sections are drawn, each view only gives the detail at that section plane and not additional background information. Fig. 8.15 gives the details at each of the section planes in a much closer and less remote arrangement.

Sections in two parallel planes

Fig. 8.16 shows a method of presenting two sections from parallel planes along the same part.

A–A

Fig. 8.16

Chapter 9
Geometrical constructions and tangency

Students will often experience difficulty in handling problems involving two and three dimensional geometrical constructions. The examples in Chapters 9 to 13 are included in order to provide a background in solving engineering problems connected with lines, planes and space. The separate chapters are grouped around applications having similar principles.

Copying a selection of these examples on the drawing board or on CAD equipment will certainly enable the reader to gain confidence. It will assist them to visualize and position the lines in space which form each part of a view, or the boundary, of a three dimensional object. It is a necessary part of draughtsmanship to be able to justify every line and dimension which appears on a drawing correctly.

Many software programs will offer facilities to perform a range of constructions, for example tangents, ellipses and irregular curves. Use these features where possible in the examples which follow.

Assume all basic dimensions where applicable.

To bisect a given angle AOB (Fig. 9.1)

1 With centre O, draw an arc to cut OA at C and OB at D.
2 With centres C and D, draw equal radii to intersect at E.
3 Line OE bisects angle AOB.

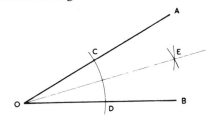

Fig. 9.1

To bisect a given straight line AB (Fig. 9.2)

1 With centre A and radius greater than half AB, describe an arc.
2 Repeat with the same radius from B, the arcs intersecting at C and D.

3 Join C to D and this line will be perpendicular to and bisect AB.

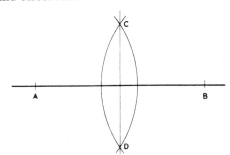

Fig. 9.2

To bisect a given arc AB (Fig. 9.3)

1 With centre A and radius greater than half AB, describe an arc.
2 Repeat with the same radius from B, the arcs intersecting at C and D.
3 Join C to D to bisect the arc AB.

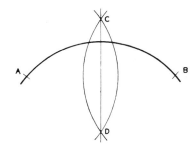

Fig. 9.3

To find the centre of a given arc AB (Fig. 9.4)

1 Draw two chords, AC and BD.
2 Bisect AC and BD as shown; the bisectors will intersect at E.
3 The centre of the arc is point E.

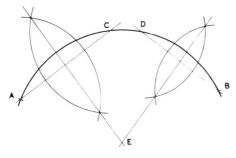

Fig. 9.4

To inscribe a circle in a given triangle ABC (Fig. 9.5)

1 Bisect any two of the angles as shown so that the bisectors intersect at D.
2 The centre of the inscribed circle is point D.

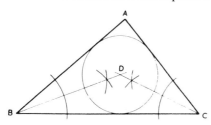

Fig. 9.5

To circumscribe a circle around triangle ABC (Fig. 9.6)

1 Bisect any two of the sides of the triangle as shown, so that the bisectors intersect at D.
2 The centre of the circumscribing circle is point D.

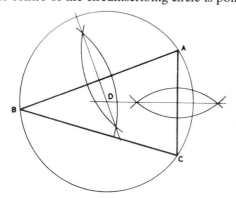

Fig. 9.6

To draw a hexagon, given the distance across the corners

Method A (Fig. 9.7(a))

1 Draw vertical and horizontal centre lines and a circle with a diameter equal to the given distance.
2 Step off the radius around the circle to give six equally spaced points, and join the points to give the required hexagon.]

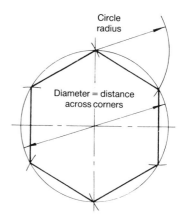

Fig. 9.7(a)

Method B (Fig. 9.7(b))

1 Draw vertical and horizontal centre lines and a circle with a diameter equal to the given distance.
2 With a 60° set-square, draw points on the circumference 60° apart.
3 Connect these six points by straight lines to give the required hexagon.

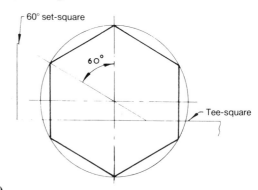

Fig. 9.7(b)

To draw a hexagon, given the distance across the flats (Fig. 9.8)

1 Draw vertical and horizontal centre lines and a circle with a diameter equal to the given distance.
2 Use a 60° set-square and tee-square as shown, to give the six sides.

Fig. 9.8

To draw a regular octagon, given the distance across corners (Fig. 9.9)

Repeat the instructions in Fig. 9.7(b) but use a 45° set-square, then connect the eight points to give the required octagon.

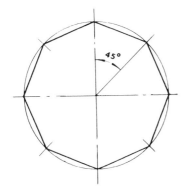

Fig. 9.9

To draw a regular octagon, given the distance across the flats (Fig. 9.10)

Repeat the instructions in Fig. 9.8 but use a 45° set-square to give the required octagon.

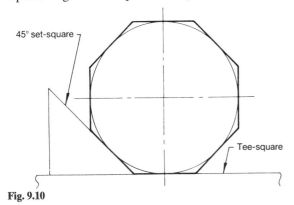

Fig. 9.10

To draw a regular polygon, given the length of the sides (Fig. 9.11)

Note that a regular polygon is defined as a plane figure which is bounded by straight lines of equal length and which contains angles of equal size. Assume the number of sides is seven in this example.

1 Draw the given length of one side AB, and with radius AB describe a semi-circle.
2 Divide the semi-circle into seven equal angles, using a protractor, and through the second division from the left join line A2.
3 Draw radial lines from A through points 3, 4, 5, and 6.
4 With radius AB and centre on point 2, describe an arc to meet the extension of line A3, shown here as point F.

5 Repeat with radius AB and centre F to meet the extension of line A4 at E.
6 Connect the points as shown, to complete the required polygon.

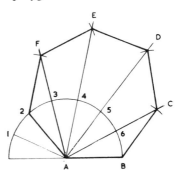

Fig. 9.11

Tangency

If a disc stands on its edge on a flat surface it will touch the surface at one point. This point is known as the point of tangency, as shown in Fig. 9.12 and the straight line which represents the flat plane is known as a tangent. A line drawn from the point of tangency to the centre of the disc is called a normal, and the tangent makes an angle of 90° with the normal.

The following constructions show the methods of drawing tangents in various circumstances.

Fig. 9.12

To draw a tangent to a point A on the circumference of a circle, centre O (Fig. 9.13)

Join OA and extend the line for a short distance. Erect a perpendicular at point A by the method shown.

Fig. 9.13

To draw a tangent to a circle from any given point A outside the circle (Fig. 9.14)

Join A to the centre of the circle O. Bisect line AO so that point B is the mid-point of AO. With centre B, draw a semi-circle to intersect the given circle at point C. Line AC is the required tangent.

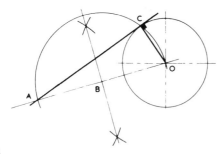

Fig. 9.14

To draw an external tangent to two circles (Fig. 9.15)

Join the centres of the circles by line AB, bisect AB, and draw a semi-circle. Position point E so that DE is equal to the radius of the smaller circle. Draw radius AE to cut the semi-circle at point G. Draw line AGH so that H lies on the circumference of the larger circle. Note that angle AGB lies in a semi-circle and will be 90° Draw line HJ parallel to BG. Line HJ will be tangential to the two circles and lines BJ and AGH are the normals.

BC = DE

Fig. 9.15

To draw an internal tangent to two circles (Fig. 9.16)

Join the centres of the circles by line AB, bisect AB and draw a semi-circle. Position point E so that DE is equal to the radius of the smaller circle BC. Draw radius AE to cut the semi-circle in H. Join AH; this line crosses the larger circle circumference at J. Draw line BH. From J draw a line parallel to BH to touch the smaller circle at K. Line JK is the required tangent. Note that angle AHB lies in a semi-circle and will therefore be 90°. AJ and BK are normals.

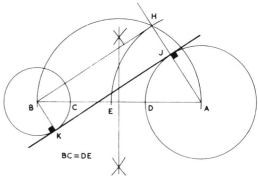

BC = DE

Fig. 9.16

To draw internal and external tangents to two circles of equal diameter (Fig. 9.17)

Join the centres of both circles by line AB. Erect perpendiculars at points A and B to touch the circumferences of the circles at points C and D. Line CD will be the external tangent. Bisect line AB to give point E, then bisect BE to give point G. With radius BG, describe a semi-circle to cut the circumference of one of the given circles at H. Join HE and extend it to touch the circumference of the other circle at J. Line HEJ is the required tangent. Note that again the angle in the semi-circle, BHE, will be 90°, and hence BH and AJ are normals.

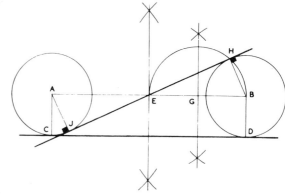

Fig. 9.17

To draw a curve of given radius to touch two circles when the circles are outside the radius (Fig. 9.18)

Assume that the radii of the given circles are 20 and 25 mm, spaced 85 mm apart, and that the radius to touch them is 40 mm.

With centre A. describe an arc equal to 20+40 = 60 mm.

With centre B, describe an arc equal to 25+40 = 65 mm.

The above arcs intersect at point C. With a radius of 40 mm, describe an arc from point C as shown, and note that the points of tangency between the arcs lie along the lines joining the centres AC and BC. It is particularly important to note the position of the points of tangency before lining in engineering drawings, so that the exact length of an arc can be established.

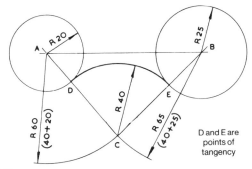

Fig. 9.18

To draw a curve of given radius to touch two circles when the circles are inside the radius (Fig. 9.19)

Assume that the radii of the given circles are 22 and 26 mm, spaced 86 mm apart, and that the radius to touch them is 100 mm.

    With centre A, describe an arc equal to $100 - 22 = 78$ mm.

    With centre B, describe an arc equal to $100 - 26 = 74$ mm.

The above arcs intersect at point C. With a radius of 100 mm, describe an arc from point C, and note that in this case the points of tangency lie along line CA extended to D and along line CB extended to E.

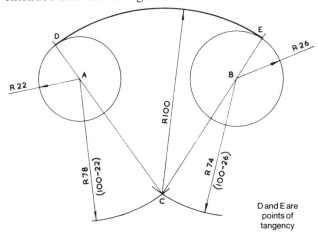

Fig. 9.19

To draw a radius to join a straight line and a given circle (Fig. 9.20)

Assume that the radius of the given circle is 20 mm and that the joining radius is 22 mm.

    With centre A, describe an arc equal to $20 + 22 = 42$ mm.

Draw a line parallel to the given straight line and at a perpendicular distance of 22 mm from it, to intersect the arc at point B.

With centre B, describe the required radius of 22 mm, and note that one point of tangency lies on the line AB at C; the other lies at point D such that BD is at 90° to the straight line.

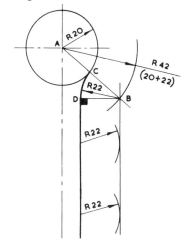

Fig. 9.20

To draw a radius which is tangential to given straight lines (Fig. 9.21)

Assume that a radius of 25 mm is required to touch the lines shown in the figures. Draw lines parallel to the given straight lines and at a perpendicular distance of 25 mm from them to intersect at points A. As above, note that the points of tangency are obtained by drawing perpendiculars through the point A to the straight lines in each case.

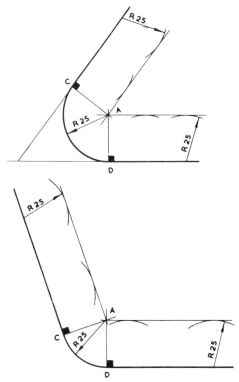

Fig. 9.21

Chapter 10
Loci applications

If a point, line, or surface moves according to a mathematically defined condition, then a curve known as a *locus* is formed. The following examples of curves and their constructions are widely used and applied in all types of engineering.

Methods of drawing an ellipse

1 Two-circle method

Construct two concentric circles equal in diameter to the major and minor axes of the required ellipse. Let these diameters be AB and CD in Fig. 10.1.

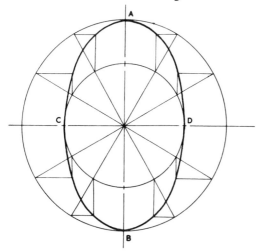

Fig. 10.1   Two-circle construction for an ellipse

Divide the circles into any number of parts; the parts do not necessarily have to be equal. The radial lines now cross the inner and outer circles.

Where the radial lines cross the outer circle, draw short lines parallel to the minor axis CD. Where the radial lines cross the inner circle, draw lines parallel to AB to intersect with those drawn from the outer circle. The points of intersection lie on the ellipse. Draw a smooth connecting curve.

2 Trammel method
Draw major and minor axes at right angles, as shown in Fig. 10.2.

Take a strip of paper for a trammel and mark on it half the major and minor axes, both measured from the same end. Let the points on the trammel be E, F, and G.

Position the trammel on the drawing so that point F always lies on the major axis AB and point G always lies on the minor axis CD. Mark the point E with each position of the trammel, and connect these points to give the required ellipse.

Note that this method relies on the difference between half the lengths of the major and minor axes, and where these axes are nearly the same in length, it is difficult to position the trammel with a high degree of accuracy. The following alternative method can be used.

Draw major and minor axes as before, but extend them in each direction as shown in Fig. 10.3.

Take a strip of paper and mark half of the major and minor axes in line, and let these points on the trammel be E, F, and G.

Position the trammel on the drawing so that point G always moves along the line containing CD; also, position point E along the line containing AB. For each position of the trammel, mark point F and join these points with a smooth curve to give the required ellipse.

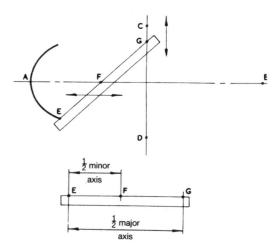

Fig. 10.2   Trammel method for ellipse construction

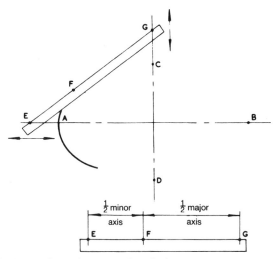

Fig. 10.3 Alternative trammel method

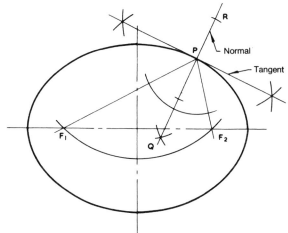

Fig. 10.5

3 To draw an ellipse using the two foci

Draw major and minor axes intersecting at point O, as shown in Fig. 10.4. Let these axes be AB and CD. With a radius equal to half the major axis AB, draw an arc from centre C to intersect AB at points F_1 and F_2. These two points are the foci. For any ellipse, the sum of the distances PF_1 and PF_2 is a constant, where P is any point on the ellipse. The sum of the distances is equal to the length of the major axis.

Divide distance OF_1 into equal parts. Three are shown here, and the points are marked G and H.

With centre F_1 and radius AG, describe an arc above and beneath line AB.

With centre F_2 and radius BG, describe an arc to intersect the above arcs.

Repeat these two steps by firstly taking radius AG from point F_2 and radius BG from F_1.

The above procedure should now be repeated using radii AH and BH. Draw a smooth curve through these points to give the ellipse.

It is often necessary to draw a tangent to a point on an ellipse. In Fig. 10.5 P is any point on the ellipse, and F_1 and F_2 are the two foci. Bisect angle F_1PF_2 with line

QPR. Erect a perpendicular to line QPR at point P, and this will be a tangent to the ellipse at point P.

The methods of drawing ellipses illustrated above are all accurate. Approximate ellipses can be constructed as follows.

Approximate method 1 Draw a rectangle with sides equal in length to the major and minor axes of the required ellipse, as shown in Fig. 10.6.

Divide the major axis into an equal number of parts; eight parts are shown here. Divide the side of the rectangle into the same equal number of parts. Draw a line from A through point 1, and let this line intersect the line joining B to point 1 at the side of the rectangle as shown. Repeat for all other points in the same manner, and the resulting points of intersection will lie on the ellipse.

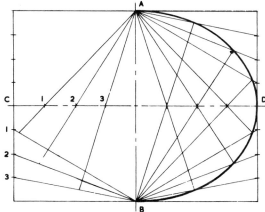

Fig. 10.6

Approximate method 2 Draw a rectangle with sides equal to the lengths of the major and minor axes, as shown in Fig. 10.7.

Bisect EC to give point F. Join AF and BE to intersect at point G. Join CG. Draw the perpendicular bisectors of lines CG and GA, and these will intersect the centre lines at points H and J.

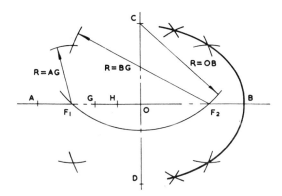

Fig. 10.4 Ellipse by foci method

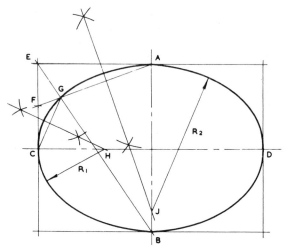

Fig. 10.7

Using radii CH and JA, the ellipse can be constructed by using four arcs of circles.

The involute

The involute is defined as the path of a point on a straight line which rolls without slip along the circumference of a cylinder. The involute curve will be required in a later chapter for the construction of gear teeth.

Involute construction

1   Draw the given base circle and divide it into, say, 12 equal divisions as shown in Fig. 10.8. Generally only the first part of the involute is required, so the given diagram shows a method using half of the length of the circumference.
2   Draw tangents at points 1, 2, 3, 4, 5 and 6.
3   From point 6, mark off a length equal to half the length of the circumference.
4   Divide line 6G into six equal parts.
5   From point 1, mark point B such that 1B is equal to one part of line 6G.
6   From point 2, mark point C such that 2C is equal to two parts of line 6G.
      Repeat the above procedure from points 3, 4 and 5, increasing the lengths along the tangents as before by one part of line 6G.
7   Join points A to G, to give the required involute.

Alternative method

1   As above, draw the given base circle, divide into, say, 12 equal divisions, and draw the tangents from points 1 to 6.
2   From point 1 and with radius equal to the chordal length from point 1 to point A, draw an arc terminating at the tangent from point 1 at point B.
3   Repeat the above procedure from point 2 with radius 2B terminating at point C.

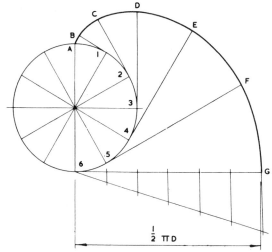

Fig. 10.8   Involute construction

4   Repeat the above instructions to obtain points D, E, F and G, and join points A to G to give the required involute.]

The alternative method given is an approximate method, but is reasonably accurate provided that the arc length is short; the difference in length between the arc and the chord introduces only a minimal error.

Archimedean spiral

The Archimedean spiral is the locus of a point which moves around a centre at uniform angular velocity and at the same time moves away from the centre at uniform linear velocity. The construction is shown in Fig. 10.9.

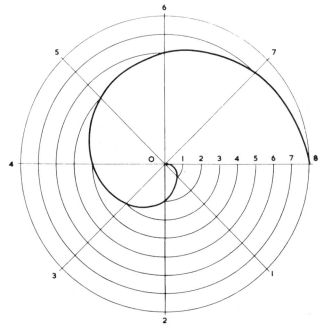

Fig. 10.9   Archimedean spiral

1 Given the diameter, divide the circle into an even number of divisions and number them.
2 Divide the radius into the same number of equal parts.
3 Draw radii as shown to intersect radial lines with corresponding numbers, and connect points of intersection to give the required spiral.

Note that the spiral need not start at the centre; it can start at any point along a radius, but the divisions must be equal.

Self-centring lathe chucks utilise Archimedean spirals.

Right-hand cylindrical helix

The helix is a curve generated on the surface of the cylinder by a point which revolves uniformly around the cylinder and at the same time either up or down its surface. The method of construction is shown in Fig. 10.10.

1 Draw the front elevation and plan views of the cylinder, and divide the plan view into a convenient number of parts (say 12) and number them as shown.
2 Project the points from the circumference of the base up to the front elevation.
3 Divide the lead into the same number of parts as the base, and number them as shown.
4 Draw lines of intersection from the lead to correspond with the projected lines from the base.
5 Join the points of intersection, to give the required cylindrical helix.
6 If a development of the cylinder is drawn, the helix will be projected as a straight line. The angle between the helix and a line drawn parallel with the base is known as the helix angle.

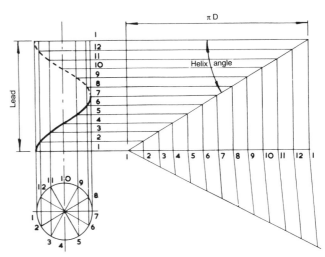

Fig. 10.10 Right-hand cylindrical helix

Note. If the numbering in the plan view is taken in the clockwise direction from point 1, then the projection in the front elevation will give a left-hand helix.

The construction for a helix is shown applied to a right-hand helical spring in Fig. 10.11. The spring is of square cross-section, and the four helices are drawn from the two outside corners and the two corners at the inside diameter. The pitch of the spring is divided into 12 equal parts, to correspond with the 12 equal divisions of the circle in the end elevation, although only half of the circle need be drawn. Points are plotted as previously shown.

A single-start square thread is illustrated in Fig. 10.12. The construction is similar to the previous problem, except that the centre is solid metal. Four helices are plotted, spaced as shown, since the thread-width is half the pitch.

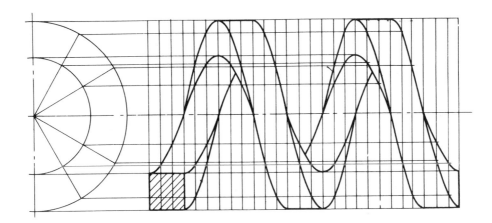

Fig. 10.11 Square-section right-hand helical spring

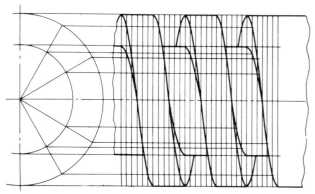

Fig. 10.12   Single-start square thread

Right-hand conical helix

The conical helix is a curve generated on the surface of the cone by a point which revolves uniformly around the cone and at the same time either up or down its surface. The method of construction is shown in Fig. 10.13.

1 Draw the front elevation and plan of the cone, and divide the plan view into a convenient number of parts (say 12) and number them as shown.
2 Project the points on the circumference of the base up to the front elevation, and continue the projected lines to the apex of the cone.

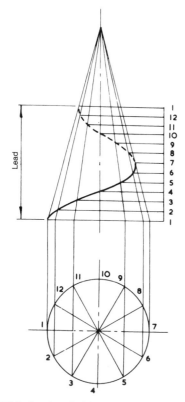

Fig. 10.13   Right-hand conical helix

3 The lead must now be divided into the same number of parts as the base, and numbered.
4 Draw lines of intersection from the lead to correspond with the projected lines from the base.
5 Join the points of intersection, to give the required conical helix.

The cycloid

The cycloid is defined as the locus of a point on the circumference of a cylinder which rolls without slip along a flat surface. The method of construction is shown in Fig. 10.14.

1 Draw the given circle, and divide into a convenient number of parts; eight divisions are shown in Fig. 10.14.
2 Divide line AA_1 into eight equal lengths. Line AA_1 is equal to the length of the circumference.
3 Draw vertical lines from points 2 to 8 to intersect with the horizontal line from centre O at points O_2, O_3, etc.
4 With radius OA and centre O_2, describe an arc to intersect with the horizontal line projected from B.
5 Repeat with radius OA from centre O_3 to intersect with the horizontal line projected from point C. Repeat this procedure.
6 Commencing at point A, join the above intersections to form the required cycloid.

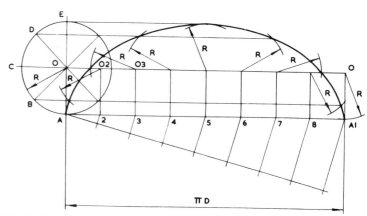

Fig. 10.14   Cycloid

The epicycloid

An epicycloid is defined as the locus of a point on the circumference of a circle which rolls without slip around the outside of another circle. The method of construction is shown in Fig. 10.15.

1 Draw the curved surface and the rolling circle, and divide the circle into a convenient number of parts (say 6) and number them as shown.

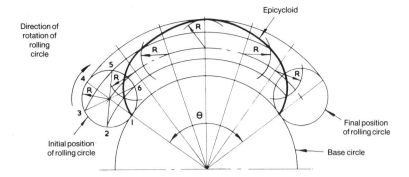

Fig. 10.15 Epicycloid

2 Calculate the length of the circumference of the smaller and the larger circle, and from this information calculate the angle θ covered by the rolling circle.

3 Divide the angle θ into the same number of parts as the rolling circle.

4 Draw the arc which is the locus of the centre of the rolling circle.

5 The lines forming the angles in step 3 will now intersect with the arc in step 4 to give six further positions of the centres of the rolling circle as it rotates.

6 From the second centre, draw radius *R* to intersect with the arc from point 2 on the rolling circle. Repeat this process for points 3, 4, 5 and 6.

7 Draw a smooth curve through the points of intersection, to give the required epicycloid.

The hypocycloid

A hypocycloid is defined as the locus of a point on the circumference of a circle which rolls without slip around the inside of another circle.

The construction for the hypocycloid (Fig. 10.16) is very similar to that for the epicycloid, but note that the rolling circle rotates in the opposite direction for this construction.

It is often necessary to study the paths taken by parts of oscillating, reciprocating, or rotating mechanisms; from a knowledge of displacement and time, informa-

tion regarding velocity and acceleration can be obtained. It may also be required to study the extreme movements of linkages, so that safety guards can be designed to protect machine operators.

Figure 10.17 shows a crank OA, a connecting rod AB, and a piston B which slides along the horizontal axis BO. P is any point along the connecting rod. To plot the locus of point P, a circle of radius OA has been divided into twelve equal parts. From each position of the crank, the connecting rod is drawn, distance AP measured, and the path taken for one revolution lined in as indicated.

The drawing also shows the piston-displacement diagram. A convenient vertical scale is drawn for the crank angle, and in this case clockwise rotation was assumed to start from the 9 o'clock position. From each position of the piston, a vertical line is drawn down to the corresponding crank-angle line, and the points of intersection are joined to give the piston-displacement diagram.

The locus of the point P can also be plotted by the trammel method indicated in Fig. 10.18. Point P_1 can be marked for any position where B_1 lies on the horizontal line, provided A_1 also lies on the circumference of the circle radius OA. This method of solving some loci problems has the advantage that an infinite number of points can easily be obtained, and these are especially useful where a change in direction in the loci curve takes place.

Figure 10.19 shows a crank OA rotating anticlockwise about centre O. A rod BC is connected to the crank at point A, and this rod slides freely through a block which is allowed to pivot at point S. The loci of points B and C are indicated after reproducing the mechanism in 12 different positions. A trammel method could also be used here if required.

Part of a shaping-machine mechanism is given in Fig. 10.20. Crank OB rotates about centre O. A is a fixed pivot point, and CA slides through the pivoting block at B. Point C moves in a circular arc of radius AC, and is connected by link CD, where point D slides horizontally. In the position shown, angle OBA is 90°, and if OB now rotates anti-clockwise at constant speed it will be seen that the forward motion of point D takes more time than the return motion. A displacement diagram for point D has been constructed as previously described.

Fig. 10.16 Hypocycloid

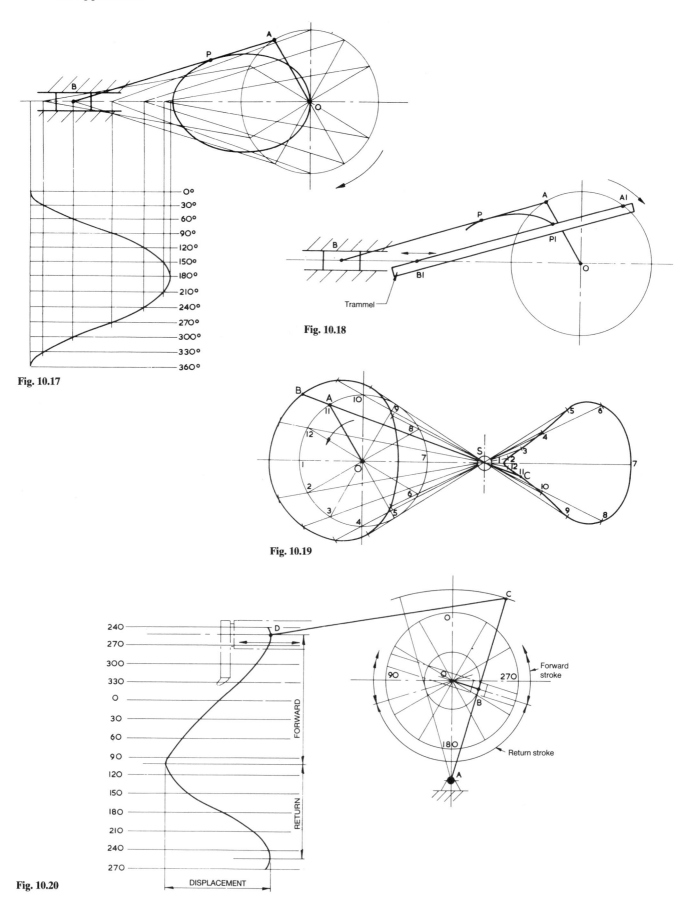

Fig. 10.17

Fig. 10.18

Trammel

Fig. 10.19

Fig. 10.20

DISPLACEMENT

Forward stroke

Return stroke

FORWARD

RETURN

Displacement diagram

Fig. 10.21

In Fig. 10.21 the radius OB has been increased, with the effect of increasing the stroke of point D. Note also that the return stroke in this condition is quicker than before.

The outlines of two gears are shown in Fig. 10.22, where the pitch circle of the larger gear is twice the pitch circle of the smaller gear. As a result, the smaller gear rotates twice while the larger gear rotates once. The mechanism has been drawn in twelve positions to plot the path of the pivot point C, where links BC and CA are connected. A trammel method cannot be applied successfully in this type of problem.

Figure 10.23 gives an example of Watt's straight-line motion. Two levers AX and BY are connected by a link AB, and the plotted curve is the locus of the mid-point

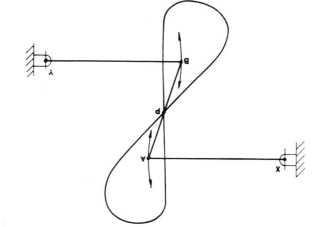

Fig. 10.23 Watt's straight line motion

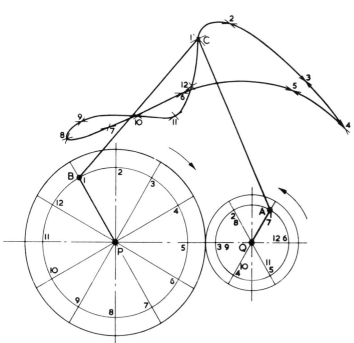

Fig. 10.22

P. The levers in this instance oscillate in circular arcs. This mechanism was used in engines designed by James Watt, the famous engineer.

A toggle action is illustrated in Fig. 10.24, where a crank rotates anticlockwise. Links AC, CD and CE are pivoted at C. D is a fixed pivot point, and E slides along the horizontal axis. The displacement diagram has been plotted as previously described, but note that, as the mechanism at E slides to the right, it is virtually stationary between points 9, 10 and 11.

The locus of any point B is also shown.

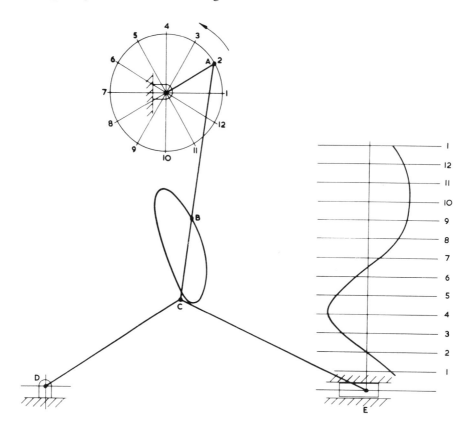

Fig. 10.24

Chapter 11
True lengths and auxiliary views

An isometric view of a rectangular block is shown in Fig. 11.1. The corners of the block are used to position a line DF in space. Three orthographic views in first-angle projection are given in Fig. 11.2, and it will be apparent that the projected length of the line DF in each of the views will be equal in length to the diagonals across each of the rectangular faces. A cross check with the isometric view will clearly show that the true length of line DF must be greater than any of the diagonals in the three orthographic views. The corners nearest to the viewing position are shown as ABCD etc.; the corners on the remote side are indicated in rings. To find the true length of DF, an auxiliary projection must be drawn, and the viewing position must be square with line DF. The first auxiliary projection in Fig. 11.2 gives the true length required, which forms part of the right-angled triangle DFG. Note that auxiliary views are drawn on planes other than the principal projection planes. A plan is projected from an elevation and an elevation from a plan. Since this is the first auxiliary view projected, and from a true plan, it is known as a *first auxiliary elevation*. Other auxiliary views could be projected from this auxiliary elevation if so required.

The true length of DF could also have been obtained by projection from the front or end elevations by

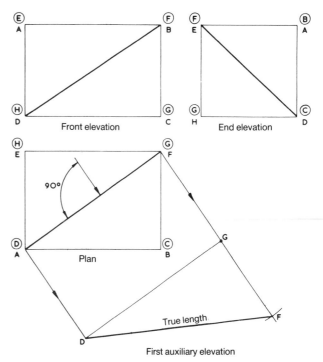

Fig. 11.2

viewing at 90° to the line, and Fig. 11.3 shows these two alternatives. The first auxiliary plan from the front elevation gives triangle FDH, and the first auxiliary plan from the end elevation gives triangle FCD, both right-angled triangles.

Figure 11.4 shows the front elevation and plan view of a box. A first auxiliary plan is drawn in the direction of arrow X. Now PQ is an imaginary datum plane at right angles to the direction of viewing; the perpendicular distance from corner A to the plane is shown as dimension 1. When the first auxiliary plan view is drawn, the box is in effect turned through 90° in the direction of arrow X, and the corner A will be situated above the plane at a perpendicular distance equal to dimension 1. The auxiliary plan view is a true view on the tilted box. If a view is now taken in the direction of arrow Y, the tilted box will be turned through 90° in the direction of the arrow, and dimension 1 to the corner will lie parallel with the plane of the paper. The other seven corners of the box are projected as indicated, and are positioned by the dimensions to the plane PQ in the

Fig. 11.1

Fig. 11.3

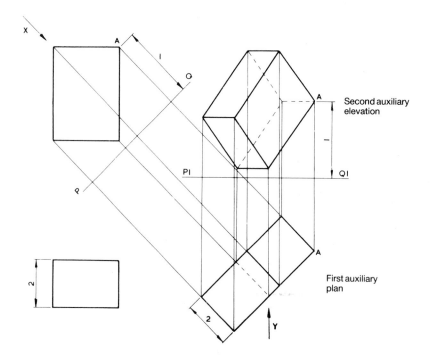

Fig. 11.4

front elevation. A match-box can be used here as a model to appreciate the position in space for each projection.

The same box has been redrawn in Fig. 11.5, but the first auxiliary elevation has been taken from the plan view in a manner similar to that described in the previous example. The second auxiliary plan projected in line with arrow Y requires dimensions from plane P_1Q_1, which are taken as before from plane PQ. Again,

check the projections shown with a match-box. All of the following examples use the principles demonstrated in these two problems.

Part of a square pyramid is shown in Fig. 11.6; the constructions for the eight corners in both auxiliary views are identical with those described for the box in Fig. 11.4.

Auxiliary projections from a cylinder are shown in Fig. 11.7; note that chordal widths in the first auxiliary

Fig. 11.5

Fig. 11.6

Fig. 11.7

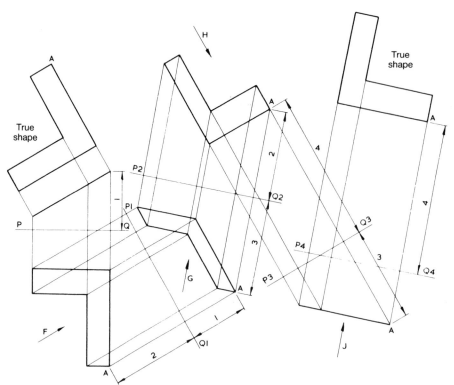

Fig. 11.8

plan are taken from the true plan. Each of twelve points around the circle is plotted in this way and then projected up to the auxiliary elevation. Distances from plane PQ are used from plane P_1Q_1. Auxiliary projections of any irregular curve can be made by plotting the positions of a succession of points from the true view and rejoining them with a curve in the auxiliary view.

Figure 11.8 shows a front elevation and plan view of a thin lamina in the shape of the letter L. The lamina lies inclined above the datum plane PQ, and the front elevation appears as a straight line. The true shape is projected above as a first auxiliary view. From the given plan view, an auxiliary elevation has been projected in line with the arrow F, and the positions of the corners above the datum plane P_1Q_1 will be the same as those above the original plane PQ. A typical dimension to the corner A has been added as dimension 1. To assist in comprehension, the true shape given could be cut from a piece of paper and positioned above the book to appreciate how the lamina is situated in space; it will then be seen that the height above the book of corner A will be dimension 2.

Now a view in the direction of arrow G parallel with the surface of the book will give the lamina shown projected above datum P_2Q_2. The object of this exercise is to show that if only two auxiliary projections are given in isolation, it is possible to draw projections to find the true shape of the component and also get the component back, parallel to the plane of the paper. The

view in direction of arrow H has been drawn and taken at 90° to the bottom edge containing corner A; the resulting view is the straight line of true length positioned below the datum plane P_3Q_3. The lamina is situated in this view in the perpendicular position above the paper, with the lower edge parallel to the paper and at a distance equal to dimension 4 from the surface. View J is now drawn square to this projected view and positioned above the datum P_4Q_4 to give the true shape of the given lamina.

In Fig. 11.9, a lamina has been made from the polygon ACBD in the development and bent along the axis AB; again, a piece of paper cut to this shape and bent to the angle ϕ may be of some assistance. The given front elevation and plan position the bent lamina in space, and this exercise is given here since every line used to form the lamina in these two views is not a true length. It will be seen that, if a view is now drawn in the direction of arrow X, which is at right angles to the bend line AB, the resulting projection will give the true length of AB, and this line will also lie parallel with the plane of the paper. By looking along the fold in the direction of arrow Y, the two corners A and B will appear coincident; also, AD and BC will appear as the true lengths of the altitudes DE and FC. The development can now be drawn, since the positions of points E and F are known along the true length of AB. The lengths of the sides AD, DB, BC and AC are obtained from the pattern development.

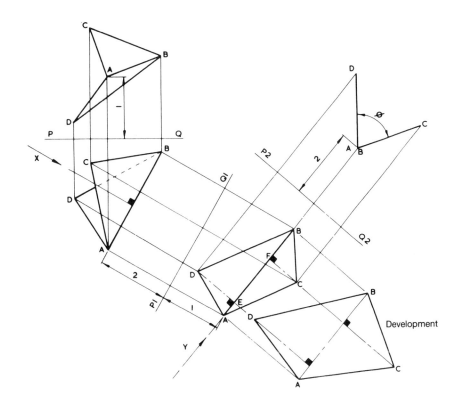

Fig. 11.9

Chapter 12
Conic sections and interpenetration of solids

Consider a right circular cone, i.e. a cone whose base is a circle and whose apex is above the centre of the base (Fig. 12.1). The true face of a section through the apex of the cone will be a triangle.

The true face of a section drawn parallel to the base will be a circle.

The true face of any other section which passes through two opposite generators will be an ellipse.

The true face of a section drawn parallel to the generator will be a parabola.

If a plane cuts the cone through the generator and the base on the same side of the cone axis, then a view on the true face of the section will be a hyperbola. The special case of a section at right-angles to the base gives a rectangular hyperbola.

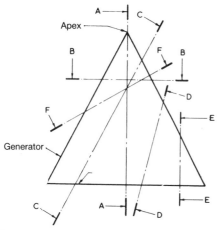

Fig. 12.1 Conic sections: section AA – triangle; section BB – circle; section CC – parabola; section DD – hyperbola; section EE – rectangular hyperbola; section FF – ellipse.

To draw an ellipse from part of a cone

Fig. 12.2 shows the method of drawing the ellipse, which is a true view on the surface marked AB of the frustum of the given cone.

1 Draw a centre line parallel to line AB as part of an auxiliary view.
2 Project points A and B onto this line and onto the centre lines of the plan and end elevation.

3 Take any horizontal section XX between A and B and draw a circle in the plan view of diameter D.
4 Project the line of section plane XX onto the end elevation.
5 Project the point of intersection of line AB and plane XX onto the plan view.
6 Mark the chord-width W on the plan, in the auxiliary view and the end elevation. These points in the auxiliary view form part of the ellipse.
7 Repeat with further horizontal sections between A and B, to complete the views as shown.

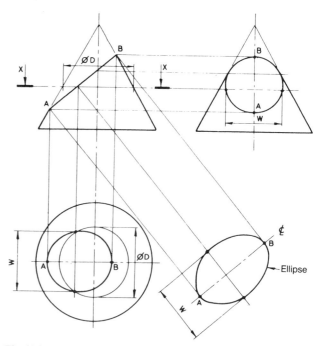

Fig. 12.2

To draw a parabola from part of a cone

Fig. 12.3 shows the method of drawing the parabola, which is a true view on the line AB drawn parallel to the sloping side of the cone.

Fig. 12.3

Fig. 12.4

1 Draw a centre line parallel to line AB as part of an auxiliary view.
2 Project point B to the circumference of the base in the plan view, to give the points B_1 and B_2. Mark chord-width B_1B_2 in the auxiliary view and in the end elevation.
3 Project point A onto the other three views.
4 Take any horizontal section XX between A and B and draw a circle in the plan view of diameter D.
5 Project the line of section plane XX onto the end elevation.
6 Project the point of intersection of line AB and plane XX to the plan view.
7 Mark the chord-width W on the plan, in the end elevation and the auxiliary view. These points in the auxiliary view form part of the parabola.
8 Repeat with further horizontal sections between A and B, to complete the three views.

To draw a rectangular hyperbola from part of a cone

Fig. 12.4 shows the method of drawing the hyperbola, which is a true view on the line AB drawn parallel to the vertical centre line of the cone.

1 Project point B to the circumference of the base in the plan view, to give the points B_1 and B_2.
2 Mark points B_1 and B_2 in the end elevation.
3 Project point A onto the end elevation. Point A lies on the centre line in the plan view.
4 Take any horizontal section XX between A and B and draw a circle of diameter D in the plan view.
5 Project the line of section XX onto the end elevation.

6 Mark the chord-width W in the plan, on the end elevation. These points in the end elevation form part of the hyperbola.
7 Repeat with further horizontal sections between A and B, to complete the hyperbola.

The ellipse, parabola, and hyperbola are also the loci of points which move in fixed ratios from a line (the directrix) and a point (the focus). The ratio is known as the *eccentricity*.

$$\text{Eccentricity} = \frac{\text{distance from focus}}{\text{perpendicular distance from directrix}}$$

The eccentricity for the ellipse is less than one.
The eccentricity for the parabola is one.
The eccentricity for the hyperbola is greater than one.

Fig. 12.5 shows an ellipse of eccentricity 3/5, a parabola of eccentricity 1, and a hyperbola of eccentricity 5/3. The distances from the focus are all radial, and the distances from the directrix are perpendicular, as shown by the illustration.

To assist in the construction of the ellipse in Fig. 12.5, the following method may be used to ensure that the two dimensions from the focus and directrix are in the same ratio. Draw triangle PA1 so that side A1 and side P1 are in the ratio of 3 units to 5 units. Extend both sides as shown. From any points B, C, D, etc., draw vertical lines to meet the horizontal at 2, 3, 4, etc.; by similar triangles, vertical lines and their corresponding horizontal lines will be in the same ratio. A similar construction for the hyperbola is shown in Fig. 12.6.

Commence the construction for the ellipse by drawing a line parallel to the directrix at a perpendicular

Fig. 12.5

(a)

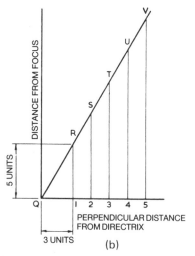

(b)

Fig. 12.6   (a) Ellipse construction (b) Hyperbola construction

distance of P3 (Fig. 12.6 (a)). Draw radius C3 from point F_1 to intersect this line. The point of intersection lies on the ellipse. Similarly, for the hyperbola (Fig. 12.6 (b)) draw a line parallel to the directrix at a perpendicular distance of Q2. Draw radius S2, and the hyperbola passes through the point of intersection. No scale is required for the parabola, as the perpendicular dis-

tances and the radii are the same magnitude.

Repeat the procedure in each case to obtain the required curves.

Interpenetration

Many objects are formed by a collection of geometrical shapes such as cubes, cones, spheres, cylinders, prisms, pyramids, etc., and where any two of these shapes meet, some sort of curve of intersection or interpenetration results. It is necessary to be able to draw these curves to complete drawings in orthographic projection or to draw patterns and developments.

The following drawings show some of the most commonly found examples of interpenetration. Basically, most curves are constructed by taking sections through the intersecting shapes, and, to keep construction lines to a minimum and hence avoid confusion, only one or two sections have been taken in arbitrary positions to show the principle involved; further similar parallel sections are then required to establish the line of the curve in its complete form. Where centre lines are offset, hidden curves will not be the same as curves directly facing the draughtsman, but the draughting principle of taking sections in the manner indicated on either side of the centre lines of the shapes involved will certainly be the same.

If two cylinders, or a cone and a cylinder, or two cones intersect each other at any angle, and the curved surfaces of both solids enclose the same sphere, then the outline of the intersection in each case will be an ellipse. In the illustrations given in Fig. 12.7 the centre

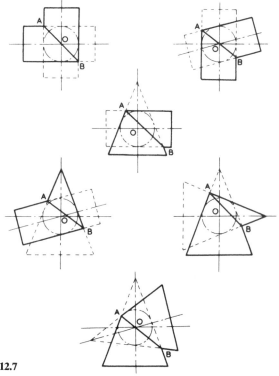

Fig. 12.7

lines of the two solids intersect at point O, and a true view along the line AB will produce an ellipse.

When cylinders of equal diameter intersect as shown in Fig. 12.8 the line at the intersection is straight and at 45°.

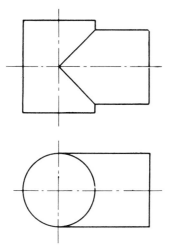

Fig. 12.8

Fig. 12.9 shows a branch cylinder square with the axis of the vertical cylinder but reduced in size. A section through any cylinder parallel with the axis produces a rectangle, in this case of width Y in the branch and width X in the vertical cylinder. Note that interpenetration occurs at points marked 3, and these points lie on a curve. The projection of the branch cylinder along the horizontal centre line gives the points marked 1, and along the vertical centre line gives the points marked 2.

Fig. 12.10 shows a cylinder with a branch on the same vertical centre line but inclined at an angle. Instead of an end elevation, the position of section AA is shown on a part auxiliary view of the branch. The construction is otherwise the same as that for Fig. 12.9.

Fig. 12.9

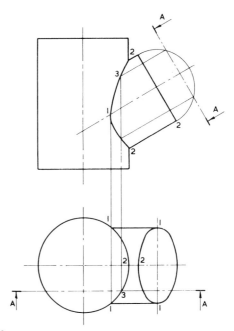

Fig. 12.10

In Fig. 12.11 the branch is offset, but the construction is similar to that shown in Fig. 12.10.

Fig. 12.12 shows the branch offset but square with the vertical axis.

Fig. 12.13 shows a cone passing through a cylinder. A horizontal section AA through the cone will give a circle of $\varnothing P$, and through the cylinder will give a rectangle of width X. The points of intersection of the circle and part of the rectangle in the plan view are projected up to the section plane in the front elevation. The plotting of more points from more sections will give the interpenetration curves shown in the front elevation and the plan.

Fig. 12.14 shows a cylinder passing through a cone. The construction shown is the same as for Fig. 12.13 in principle.

Fig. 12.15 shows a cone and a square prism where interpenetration starts along the horizontal section BB at point 1 on the smallest diameter circle to touch the prism. Section AA is an arbitrary section where the projected diameter of the cone $\varnothing X$ cuts the prism in the plan view at the points marked 2. These points are then projected back to the section plane in the front elevation and lie on the curve required. The circle at section CC is the largest circle which will touch the prism across the diagonals in the plan view. Having drawn the circle in the plan view, it is projected up to the sides of the cone in the front elevation, and points 3 at the corners of the prism are the lowest points of contact.

A casting with a rectangular base and a circular-section shaft is given in Fig. 12.16. The machining of the radius $R1$ in conjunction with the milling of the flat surfaces produces the curve shown in the front elevation. Point 1 is shown projected from the end elevation. Section AA produces a circle of $\varnothing X$ in the plan view

Fig. 12.11

Fig. 12.12

Fig. 12.13

Fig. 12.14

Fig. 12.15

Fig. 12.16

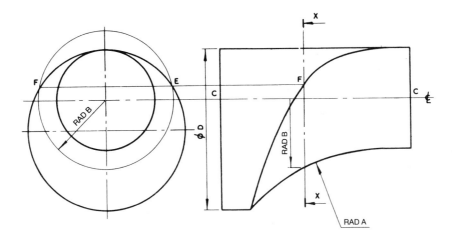

Fig. 12.17

and cuts the face of the casting at points marked 2, which are transferred back to the section plane. Similarly, section BB gives $\varnothing Y$ and points marked 3. Sections can be taken until the circle in the plan view increases in size to $R2$; at this point, the interpenetration curve joins a horizontal line to the corner of the casting in the front elevation.

In Fig. 12.17 a circular bar of diameter D has been turned about the centre line CC and machined with a radius shown as RAD A. The resulting interpenetration curve is obtained by taking sections similar to section XX. At this section plane, a circle of radius B is projected in the front elevation and cuts the circum-

ference of the bar at points E and F. The projection of point F along the section plane XX is one point on the curve. By taking a succession of sections, and repeating the process described, the curve can be plotted.

Note that, in all these types of problem, it rarely helps to take dozens of sections and then draw all the circles before plotting the points, as the only result is possible confusion. It is recommended that one section be taken at a time, the first roughly near the centre of any curve, and others sufficiently far apart for clarity but near enough to maintain accuracy. More sections are generally required where curves suddenly change direction.

Chapter 13
Development of patterns from sheet materials

Many articles such as cans, pipes, elbows, boxes, ducting, hoppers, etc. are manufactured from thin sheet materials. Generally a template is produced from an orthographic drawing when small quantities are required (larger quantities may justify the use of press tools), and the template will include allowances for bending and seams, bearing in mind the thickness of material used.

Exposed edges which may be dangerous can be wired or folded, and these processes also give added strength, e.g. cooking tins and pans. Some cooking tins are also formed by pressing hollows into a flat sheet. This type of deformation is not considered in this chapter, which deals with bending or forming in one plane only. Some common methods of finishing edges, seams, and corners are shown in Fig. 13.1.

Fig. 13.1

Fig. 13.2

The following examples illustrate some of the more commonly used methods of development in patternmaking, but note that, apart from in the first case, no allowance has been made for joints and seams.

Where a component has its surfaces on flat planes of projection, and all the sides and corners shown are true lengths, the pattern is obtained by parallel-line or straight-line development. A simple application is given in Fig. 13.2 for an open box.

The development of a hexagonal prism is shown in Fig. 13.3. The pattern length is obtained by plotting the distances across the flat faces. The height at each corner is projected from the front elevation, and the top of the prism is drawn from a true view in the direction of arrow X.

An elbow joint is shown developed in Fig. 13.4. The length of the circumference has been calculated and

Fig. 13.3

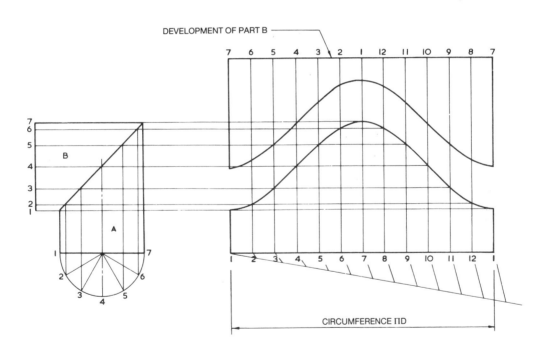

Fig. 13.4

divided into twelve equal parts. A part plan, divided into six parts, has the division lines projected up to the joint, then across to the appropriate point on the pattern. It is normal practice on a development drawing to leave the joint along the shortest edge; however, on part B the pattern can be cut more economically if the joint on this half is turned through 180°.

An elbow joint made from four parts has been completely developed in Fig. 13.5. Again, by alternating the position of the seams, the patterns can be cut with no waste. Note that the centre lines of the parts marked B and C are 30° apart, and that the inner and outer edges are tangential to the radii which position the elbow.

A thin lamina is shown in orthographic projection in Fig. 13.6. The development has been drawn in line with

Fig. 13.5

Fig. 13.6

the plan view by taking the length along the front elevation in small increments of width *C* and plotting the corresponding depths from the plan.

A typical interpenetration curve is given in Fig. 13.7. The development of part of the cylindrical portion is shown viewed from the inside. The chordal distances on the inverted plan have been plotted on either side of the centre line of the hole, and the corresponding heights have been projected from the front elevation. The method of drawing a pattern for the branch is identical to that shown for the two-piece elbow in Fig. 13.4.

An example of radial-line development is given in Fig. 13.8. The dimensions required to make the development are the circumference of the base and the slant height of the cone. The chordal distances from the plan view have been used to mark the length of arc required for the pattern; alternatively, for a higher degree of accuracy, the angle can be calculated and then subdivided. In the front elevation, lines O1 and O7 are true

lengths, and distances OG and OA have been plotted directly onto the pattern. The lines O2 to O6 inclusive are not true lengths, and, where these lines cross the sloping face on the top of the conical frustum, horizontal lines have been projected to the side of the cone and been marked B, C, D, E, and F. True lengths OF, OE, OD, OC, and OB are then marked on the pattern. This procedure is repeated for the other half of the cone. The view on the sloping face will be an ellipse, and the method of projection has been described in Chapter 12.

Part of a square pyramid is illustrated in Fig. 13.9. The pattern is formed by drawing an arc of radius OA and stepping off around the curve the lengths of the base, joining the points obtained to the apex O. Distances OE and OG are true lengths from the front elevation, and distances OH and OF are true lengths from the end elevation. The true view in direction of arrow X completes the development.

The development of part of a hexagonal pyramid is

Fig. 13.7

Fig. 13.8

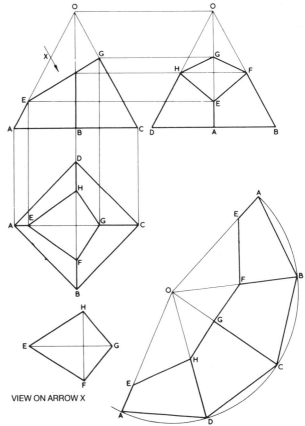

VIEW ON ARROW X

Fig. 13.9

shown in Fig. 13.10. The method is very similar to that given in the previous example, but note that lines OB, OC, OD, OE, and OF are true lengths obtained by projection from the elevation.

Fig. 13.11 shows an oblique cone which is developed by triangulation, where the surface is assumed to be formed from a series of triangular shapes. The base of the cone is divided into a convenient number of parts (12 in this case) numbered 0–6 and projected to the front elevation with lines drawn up to the apex A. Lines 0A and 6A are true-length lines, but the other five shown all slope at an angle to the plane of the paper. The true lengths of lines 1A, 2A, 3A, 4A, and 5A are all

equal to the hypotenuse of right-angled triangles where the height is the projection of the cone height and the base is obtained from the part plan view by projecting distances B1, B2, B3, B4, and B5 as indicated.

Assuming that the join will be made along the

Fig. 13.10

Fig. 13.11

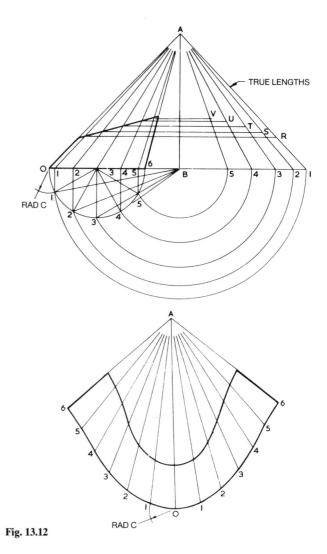

Fig. 13.12

shortest edge, the pattern is formed as follows. Start by drawing line 6A, then from A draw an arc on either side of the line equal in length to the true length 5A. From point 6 on the pattern, draw an arc equal to the chordal distance between successive points on the plan view. This curve will intersect the first arc twice at the points marked 5. Repeat by taking the true length of line 4A and swinging another arc from point A to intersect with chordal arcs from points 5. This process is continued as shown on the solution.

Fig. 13.12 shows the development of part of an oblique cone where the procedure described above is followed. The points of intersection of the top of the cone with lines 1A, 2A, 3A, 4A, and 5A are transferred to the appropriate true-length constructions, and true-length distances from the apex A are marked on the pattern drawing.

A plan and front elevation is given in Fig. 13.13 of a transition piece which is formed from two halves of oblique cylinders and two connecting triangles. The plan view of the base is divided into 12 equal divisions, the sides at the top into 6 parts each. Each division at the bottom of the front elevation is linked with a line to

Fig. 13.13

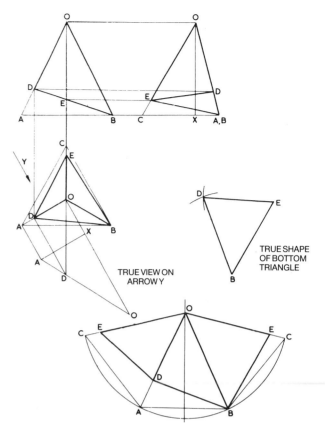

Fig. 13.14

the similar division at the top. These lines, P1, Q2, etc., are all the same length. Commence the pattern construction by drawing line S4 parallel to the component. Project lines from points 3 and R, and let these lines intersect with arcs equal to the chordal distances *C*, from the plan view, taken from points 4 and S. Repeat the process and note the effect that curvature has on the distances between the lines projected from points P, Q, R, and S. After completing the pattern to line P1, the triangle is added by swinging an arc equal to the length *B* from point P, which intersects with the arc shown, radius *A*. This construction for part of the pattern is continued as indicated.

Part of a triangular prism is shown in Fig. 13.14, in orthographic projection. The sides of the prism are constructed from a circular arc of true radius OC in the end elevation. Note that radius OC is the only true length of a sloping side in any of the three views. The base length CA is marked around the circumference of the arc three times, to obtain points A, B, and C. True length OE can be taken from the end elevation, but a construction is required to find the true length of OD. Draw an auxiliary view in direction with arrow Y, which is square to line OA as shown. The height of the triangle, OX, can be taken from the end elevation. The projection of point D on the side of the triangle gives the true length OD. The true shape at the bottom can be drawn by taking lengths ED, DB, and BE from the pattern and constructing the triangle shown.

A transition piece connecting two rectangular ducts is given in Fig. 13.15. The development is commenced by drawing the figure CBFG, and the centre line of this part can be obtained from the front elevation which appears as line CG, the widths being taken from the plan. The next problem is to obtain the true lengths of lines CG and DH and position them on the pattern; this can be done easily by the construction of two triangles, after the insertion of line DG. The true lengths can be found by drawing right-angled triangles where the base measurements are indicated as dimensions 1, 2, and 3, and the height is equal to the height of the front

Fig. 13.15

Fig. 13.16

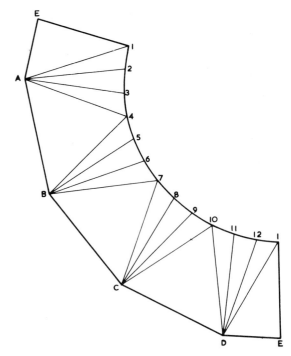

elevation. The length of the hypotenuse in each case is used as the radius of an arc to form triangles CDG and GDH. The connecting seam is taken along the centre line of figure ADHE and is marked JK. The true length of line JK appears as line HD in the front elevation, and the true shape of this end panel has been drawn beside the end elevation to establish the true lengths of the dotted lines EK and HK, since these are used on the pattern to draw triangles fixing the exact position of points K and J.

A transition piece connecting square and circular ducts is shown in Fig. 13.16. The circle is divided into twelve equal divisions, and triangles are formed on the surface of the component as shown. A construction is required to establish the true lengths of lines A1, A2, A3, and A4. These lengths are taken from the hypotenuse of right-angled triangles whose height is equal to the height of the front elevation, and the base measurement is taken from the projected lengths in the plan view. Note that the lengths A2 and A3 are the same, as are A1 and A4, since the circle lies at the centre of the square in the plan. The constructions from the other three corners are identical to those from corner A. To form the pattern, draw a line AB, and from A describe an arc of radius A4. Repeat from end B, and join the triangle. From point 4, swing an arc equal to the chordal length between points 4 and 3 in the plan view, and let this arc intersect with the true length A3, used as a radius from point A. Mark the intersection as point 3. This process is repeated to form the pattern shown. The true length of the seam at point E can be measured from the front elevation. Note that, although chordal distances are struck between successive points around the pattern, the points are themselves joined by a curve; hence no ultimate error of any significance occurs when using this method.

Fig. 13.17 shows a similar transition piece where the top and bottom surfaces are not parallel. The construction is generally very much the same as described above, but two separate true-length constructions are required for the corners marked AD and BC. Note

that, in the formation of the pattern, the true length of lines AB and CD is taken from the front elevation when triangles AB4 and DC10 are formed. The true length of the seam is also the same as line A1 in the front elevation.

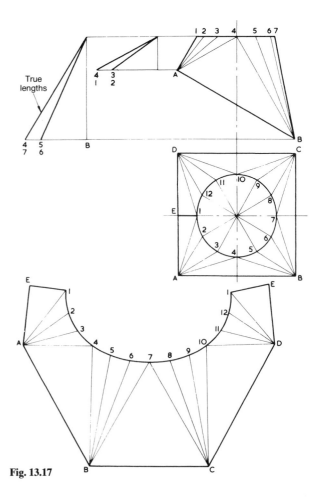

Fig. 13.17

Chapter 14
Dimensioning principles

A drawing should provide a complete set of working instructions for the craftsmen who produce the component. Dimensions define geometric characteristics such as angles, diameters, lengths, and positions; and each dimension which defines a characteristic should appear on the drawing only once, and it should not be necessary for the craftsman either to scale the drawing or to deduce dimensions by the subtraction or addition of other dimensions.

Theoretically any component can be analysed and divided into a number of standard common geometrical shapes such as cubes, prisms, cylinders, parts of cones, etc. A hole through a plate can be considered as a cylinder for this purpose. Dimensioning is a means of defining the exact amount of material remaining after a series of manufacturing operations have taken place. Manufacturing is followed by inspection and measurement is taken in both cases from functional axes or planes and these are known as datums.

A solid block with a hole in it is shown in Fig. 14.1 and to establish the exact shape of the item we require to know the dimensions which govern its length, height and thickness also the diameter and depth of the hole and its position in relation to the surface of the block.

The axis of the hole is shown at the intersection of two centre lines positioned from the left hand side and the bottom of the block and these two surfaces have been taken as datums. The length and height have also been measured from these surfaces separately and this is a very important point as errors may become cumulative and this is discussed later in the chapter.

Dimensioning therefore, should be undertaken with a view to defining the shape or form and overall size of the component carefully, also the sizes and positions of the various features, such as holes, counterbores, tappings etc., from the necessary datum planes or axes.

The completed engineering drawing should also include sufficient information for the manufacture of the part and this involves the addition of notes regarding the materials used, tolerances of size, limits and fits, surface finishes, the number of parts required and any further comments which result from a consideration of the use to which the completed component will be put. For example, the part could be used in sub-assembly and notes would then make reference to associated drawings or general assemblies.

Now there are a few rules applicable to dimensioning which are covered by British Standard 308 and, if these are adhered to, it is reasonably easy to produce a drawing to a good professional standard.

1 Dimension and projection lines are thin continuous lines 0.35 mm thick, if possible, clearly placed outside the outline of the drawing. As previously mentioned, the drawing outline should be lined in with a thickness of 0.7 mm and the drawing outline will then be clearly defined and in contrast with the dimensioning system.
2 The projection lines should not touch the drawing but a small gap should be left, about 2 to 3 mm, depending on the size of the drawing. The projection lines should then continue for the same distance past the dimension line.
3 Arrowheads should be approximately triangular, must be of uniform size and shape and in every case touch the dimension line to which they refer Arrowheads drawn manually should be filled in. Arrowheads drawn by machine need not be filled in.
4 Bearing in mind the size of the actual dimensions and the fact that there may be two numbers together where limits of size are quoted, then adequate space must be left between rows of dimensions and a spacing of about 12 mm is recommended.

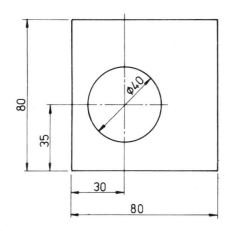

Fig. 14.1

5   Centre lines must never be used as dimension lines but must be left clear and distinct. They can be extended, however, when used in the role of projection lines.

6   Dimensions are quoted in millimetres to the minimum number of significant figures. For example, 19 and not 19.0. In the case of a decimal dimension, we always use a naught before the decimal marker which might not be noticed on a drawing print which has poor line definition. We write 0.4 and not .4. It should be stated here that on metric drawings the decimal marker may be either a full stop or a comma positioned on the base line between the figures, for example, 5.2 or 5,2 but never 5·2 with a decimal point midway.

7   To enable dimensions to be read clearly, figures are placed so that they can be read from the bottom of the drawing, or by turning the drawing in a clockwise direction, so that they can be read from the right hand side.

8   Leader lines are used to indicate where specific notes apply. The leader line to the hole is directed towards the centre point but terminates at the circumference in an arrow. A leader line for a part number terminates in a dot within the outline of the component. The gauge plate here is assumed to be part number six of a set of inspection gauges.

Fig. 14.2 shows a partly completed drawing of a gauge to illustrate the above aspects of dimensioning.

When components are drawn in orthographic projection, a choice often exists where to place the dimensions and the following general rules will give assistance.

1   Start by dimensioning the view which gives the clearest understanding of the profile or shape of the component.

2   If space permits, and obviously this varies with the size and degree of complexity of the subject, place the dimensions outside the profile of the component as first choice.

3   Where several dimensions are placed on the same side of the drawing, position the shortest dimension nearest to the component and this will avoid dimension lines crossing.

4   Try and ensure that similar spacings are made between dimension lines as this gives a neat appearance on the completed drawing.

5   Overall dimensions which are given for surfaces that can be seen in two projected views are generally best positioned between these two views.

Remember, that drawings are used for the manufacture of components, so ask yourself whether or not the machinist or fitter can use or work to the dimension you have quoted to make the item. Also, can the inspector check the figure, in other words, is it a measurable distance?

Fig. 14.3 shows a component which has been partly dimensioned to illustrate some of the principles involved.

Careless and untidy dimensioning can spoil an otherwise sound drawing and it should be stated that many marks are lost in examinations due to poor quality work.

Fig. 14.3

Fig. 14.2   Gauge plate

Dimensions not to scale

Sometimes it is desirable to dimension part of a drawing which has not been drawn to the correct scale. The particular dimension is then underlined (Fig. 14.4)

Fig. 14.4

Chain dimensioning

A plan view of a twist drill stand is given in Fig. 14.5 to illustrate chain dimensioning.

Now each of the dimensions in the chain would be subject to a manufacturing tolerance since it is not possible to mark out and drill each of the centre distances exactly. As a test of drawing accuracy, start at the left hand side and mark out the dimensions shown in turn. Measure the overall figure on your drawing and check with the auxiliary dimension given. Note the considerable variation in length which results from small errors in each of the six separate dimensions in the chain which clearly accumulate. Imagine the effect of marking out say twenty holes for rivets in each of two plates, how many holes would eventually line up? The overall length is shown in parentheses (157) and is known as an auxiliary dimension. This dimension is not one which is worked to in practice but is given purely for reference purposes. You will now appreciate that it will depend on the accuracy with which each of the pitches in the chain is marked out.

Fig. 14.5

Parallel dimensioning

Improved positional accuracy is obtainable by dimensioning from a datum and this method is shown in Fig. 14.6. The datum selected is the left hand side of the stand. Note that the overall length is not now shown as an auxiliary dimension, but as a fixed distance which

Fig. 14.6

can be separately marked out, manufactured and definitely checked.

Simplified parallel dimensioning

Note that the spacing of the six dimensions takes up considerable space and this space can be reduced if the following convention is adopted. The common origin is indicated as shown with a thin continuous circle and the dimensions placed near the arrowheads.

Fig. 14.7

Staggered dimensions

A number of parallel dimensions may be staggered as shown for clarity (Fig. 14.8).

The dimension lines can sometimes be omitted and leader lines used in their place, these are positioned at right angles to the outline of the component (Fig. 14.9).

Fig. 14.8

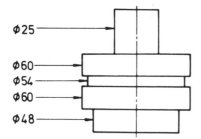

Fig. 14.9

Dimensioning circles

The symbol ⌀ preceding the figure is used for specifying diameters and it should be written as large as the figures which establish the size, e.g. ⌀065. Alternative methods of dimensioning diameters are given below. The size of hole and space available on the drawing generally dictates which method the draughtsman choses.

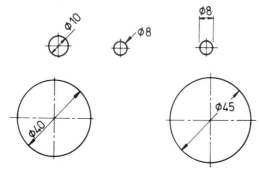

Fig. 14.10

Dimensioning radii

Alternative methods are shown where the position of the centre of the arc need not be located. Note that the dimension line is drawn through the arc centre or lies in a line with it in the case of short distances and the arrowhead touches the arc.

Fig. 14.11

Dimensioning spherical radii and diameters

Fig. 14.12

Dimensioning curves

A curve formed by the blending of several radii must have the radii with their centres of curvature clearly marked.

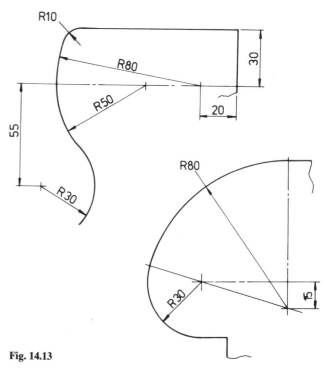

Fig. 14.13

Dimensioning irregular curves

Irregular curves may be dimensioned by the use of ordinates. To illustrate the use of ordinates, a section through the hull of a boat is shown (Fig. 14.14). Since the hull is symmetrical about the vertical centre line it is not necessary to draw both halves in full and if the curve is presented in this manner then two short thick parallel lines are drawn at each end of the profile at right angles to the centre line. The outline is also extended slightly beyond the centre line to indicate that the shape is to be continued. Ordinates are then positioned on the drawing and the outline passes through each of the chosen fixed points.

Fig. 14.14

Unidirectional and aligned dimensions

Both methods are in common use.

1 Unidirectional dimensions are drawn parallel with the bottom of the drawing sheet, also any notes which refer to the drawing use this method.

Fig. 14.15

2 Aligned dimensions are shown in parallel with the related dimension line and positioned so that they can be read from the bottom of the drawing or from the right hand side (Fig. 14.16).

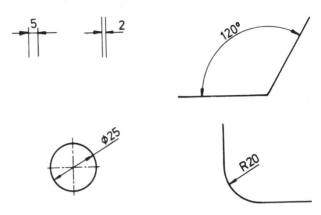

Fig. 14.16

Angular dimensions

Angular dimensions on engineering drawings are expressed as follows:

a) Degrees, e.g. 30°.
b) Degrees and minutes, e.g. 30° 40'.
c) Degrees, minutes and seconds e.g. 30° 40' 20".

For clarity a full space is left between the degree symbol and the minute figure also between the minute symbol and the second figure.

In the case of an angle less than one degree it should be preceded by 0°, e.g. 0° 25'.

Fig. 14.17 shows various methods of dimensioning angles.

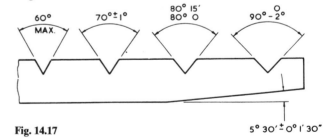

Fig. 14.17

Tapers

In Fig. 14.18 the difference in magnitude between dimensions X and Y (whether diameters or widths) divided by the length between them defines a ratio known as a *taper*.

$$\text{Taper} = \frac{X - Y}{\text{length}} \equiv 2 \tan \frac{\theta}{2}$$

For example, the conical taper in Fig. 14.19

$$= \frac{20 - 10}{40} = \frac{10}{40} = 0.25$$

Fig. 14.18

and may be expressed as rate of taper 0.25:1 on diameter.

The ISO recommended symbol for taper is ➤, and this symbol can be shown on drawings accompanying the rate of taper,

i.e. ➤ 0.25:1

The arrow indicates the direction of taper.

When a taper is required as a datum, it is enclosed in a box as follows:

Fig. 14.19

Dimensioning tapers

The size, form, and position of a tapered feature can be defined by calling for a suitable combination of the following:

1 the rate of taper, or the included angle;
2 the diameter or width at the larger end;
3 the diameter or width at the smaller end;
4 the length of the tapered feature;
5 the diameter or width at a particular cross-section, which may lie within or outside the feature concerned;
6 the locating dimension from the datum to the cross-section referred to above.

Care must be taken to ensure that no more dimensions are quoted on the drawing than are necessary. If reference dimensions are given to improve communications, then they must be shown in brackets, e.g. (1:5 taper).

Fig. 14.20 gives four examples of the methods used to specify the size, form, and position of tapered features.

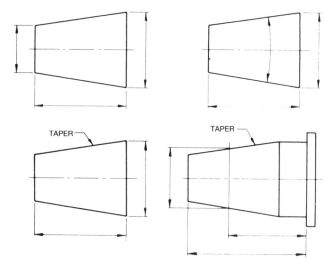

Fig. 14.20

Dimensioning two mating tapers

This method is used when the fit to a mating component or gauge is necessary; the following information should then be added to the feature:

1 'TO FIT PART NO. YYY',
2 'TO FIT GAUGE (PART NO. GGG)'.

When note 2 is added to the drawing, this implies that a 'standard rubbing gauge' will give an acceptable even marking when 'blued'. The functional requirement, whether the end-wise location is important or not, will determine the method and choice of dimensioning.

An example of dimensioning two mating tapers when end-wise location is important, and adjacent dimensions are unimportant, is shown in Fig. 14.21. Note that the datum diameter could be inside a female cone.

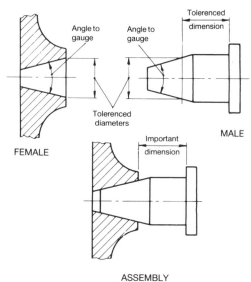

Fig. 14.21

Dimensioning chamfers

Alternative methods of dimensioning internal and external chamfers are shown in Fig. 14.22.

Fig. 14.22

Dimensioning squares or flats

Fig. 14.23 shows a square machined on the end of a shaft so that it can be turned by means of a spanner. The diagonal lines are added to indicate the flat surface.

Fig. 14.23

Part of a spindle which carries the chain wheel of a cycle, secured by a cotter pin, illustrates a flat surface which is not at the end of the shaft (Fig. 14.24).

Fig. 14.24

Dimensioning holes

The depth of drilled holes, when stated in note form, refers to the depth of the cylindrical portion and not to the point left by the drill. If no other indication is given they are assumed to go through the material. Holes in flanges or bosses are generally positioned around a pitch circle (PCD) and may be spaced on the main

centre lines of the component (on centres) or as shown below equally spaced off centres. Holes are usually drilled off centres to provide for maximum access to fixing bolts in the case of valves and pipeline fittings. Special flanges need to have each hole positioned individually and an example is given with three tapped holes (see Fig. 14.25).

Fig. 14.25

Dimensioning counterbores

A drilling machine is used for this operation, and a typical counterboring tool is shown in Fig. 14.26. The operation involves enlarging existing holes, and the depth of the enlarged hole is controlled by a stop on the drilling machine. The location of the counterbored hole

Fig. 14.26

is assisted by a pilot at the tip of the tool which is a clearance fit in the previously drilled hole. A typical use for a counterbored hole is to provide a recess for the head of a screw, as shown in Fig. 14.27 or a flat surface

Fig. 14.27

for an exposed nut or bolt, as in Fig. 14.28. The flat surface in Fig. 14.28 could also be obtained by spotfacing.

FLAT SURFACE ON CASTING OBTAINED BY COUNTERBORING OR SPOTFACING

Fig. 14.28

Fig. 14.29 shows methods of dimensioning counterbores. Note that, in every case, it is necessary to specify the size of counterbore required. It is not sufficient to state 'COUNTERBORE FOR M10 RD HD SCREW', since obviously the head of the screw will fit into any counterbore which is larger than the head.

Dimensioning countersunk holes

Countersinking is also carried out on a drilling machine, and Fig. 14.30 shows typical tools. Included angles of 60° and 90° are commonly machined, to accommodate the heads of screws and rivets to provide a flush finish.

Fig. 14.29

Fig. 14.30(a)   Taper-shank countersink (with 60° or 90° included angle of countersink)

Fig. 14.30(b)   Straight-shank machine countersink (with 60° or 90° included angle of countersink)

Note. Refer to manufacturers' catalogues for dimensions of suitable rivets and screws.

Fig. 14.31

Dimensioning spotfaces

Spotfacing is a similar operation to counterboring, but in this case the metal removed by the tool is much less. The process is regularly used on the surface of castings, to provide a flat seating for fixing bolts. A spotfacing tool is shown in Fig. 14.32, where a loose cutter is used. The length of cutter controls the diameter of the spotface. As in the counterboring operation, the hole must be previously drilled, and the pilot at the tip of the spotfacing tool assists in location.

Fig. 14.32

Fig. 14.33 shows the method of dimensioning. Note that, in both cases, the depth of spotface is just sufficient to remove the rough surface of the casting over the 40 mm diameter area.

Fig. 14.33

Dimensioning for manufacture

It should be emphasized that dimensioning must be performed with the user of the drawing very much in mind. In the case of the finished bearing housing shown in Fig. 14.35 two different production processes are involved in its manufacture; namely casting and machining of the component. It is sometimes preferable to produce two separate drawings, one to show the dimensions of the finished casting and the other to show the dimensions which are applicable to the actual machining operation. Fig. 14.34 shows a suitable drawing for the casting patternmaker. Allowances are made for machining and also for the fact that the casting will shrink when it cools. The machinist will take the rough casting and remove metal to produce the finished component, all other surfaces having a rough finish. Fig. 14.35 shows the required dimensions for machining. Note that the bore of the casting is required to be finished between the two sizes quoted for functional purposes.

Fig. 14.34

Fig. 14.35

Machining and surface-roughness symbols

The quality of any finished surface has a direct connection with the function and wear of the component. Surface finishes vary considerably in quality, and the maximum roughness acceptable is quoted only once on the surface, near to the dimension fixing the location or size of the surface.

The proportions of typical symbols are shown in Fig. 14.36. The surface finish quoted on a detail drawing will, unless a note is added to the contrary, be the final finish required for the correct functioning of the part in service. The component may undergo several stages in manufacture, for example turning then grinding or chrome-plating after pressing; the roughness at the intermediate stage is normally not quoted. If only machining is required to be indicated, without reference to roughness, then the number (25 in the example above) is omitted.

Fig. 14.36

Fig. 14.37 Maximum and minimum values of surface roughness

In certain circumstances it may be advantageous to quote surface roughness within an acceptable range of maximum and minimum values. This is done by quoting the maximum limit above the min. limit as shown on the machined example, Fig. 14.37.

Certain components are required to be machined to the same quality of finish on every surface, and, to avoid the repeated placing of the symbol on the drawing, a single general note may be added, as in Fig. 14.38.

Fig. 14.38 Symbol denoting 'MACHINE ALL OVER'

Where a specific process or production method is required to be used to obtain a certain finish, then the machining symbol is slightly modified to make a reference to the fact. The horizontal line is added and the process is indicated as in Fig. 14.39.

Fig. 14.39

Finished surfaces are not only obtained by machining; components with acceptable finished surfaces can be manufactured by die casting and pressing, for exam-

ple. If a control is required on the quality of a surface that does not necessarily need machining, then the horizontal line on the symbol is omitted (Fig. 14.40).

Fig. 14.40 Symbol where machining is not mandatory

In some applications friction, for example, may be desirable, and the machining of a component already produced by another process is definitely not required. The letter 'O' is then inserted in the vee of the machining symbol, to signify the omission, as in Fig. 14.41; alternatively a note 'DO NOT MACHINE' may be used.

Fig. 14.41 Symbol to indicate that removal of material is not permitted in order to achieve the surface roughness quoted

The marks of machining are inevitable but not always undesirable; they can be used to assist in the correct functioning of parts. These marks are a succession of minute hills and valleys, and mating parts can be arranged, if necessary, with machining marks in line with one another or at other angles. Machining marks can also be used as a form of decoration on finished surfaces. If this special requirement is desired, the direction of lay must be indicated in association with the machining symbol. The end of the shaft in Fig. 14.42 will be grooved like a gramophone record. The surface of the block in Fig. 14.43 will have minute grooves parallel with the line of the arrow.

Fig. 14.42

Fig. 14.43

Fig. 14.44 shows approximate surface-roughness ranges for components manufactured by some common production methods. This information can only be approximate, since finish depends on many factors, such as the skill of the machinist, the accuracy and condition of the machine, the speeds and feeds selected

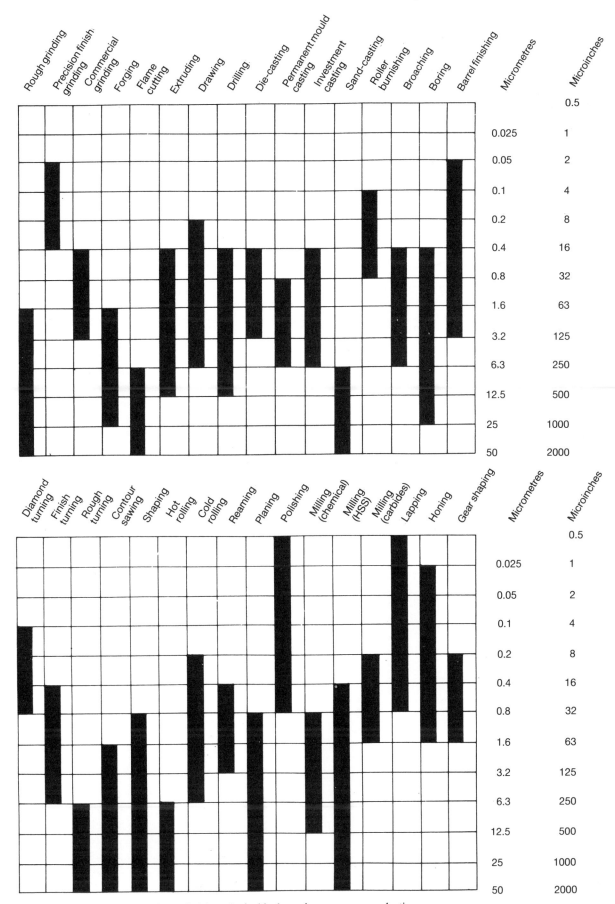

Fig. 14.44 Approximate surface roughness heights obtainable by various common production processes

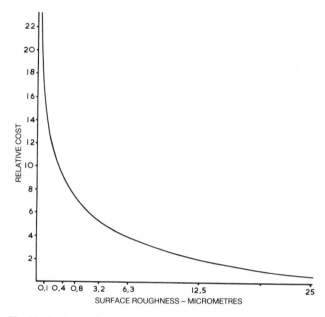

Fig. 14.45 Approximate relationship between surface roughness and cost.

for the operation, and the quality and condition of the cutting tools.

The approximate relationship between surface roughness and the cost of producing such a finish is shown in Fig. 14.45. The cost of rough machining can be considered as the zero datum on the y axis of the graph, and other processes can be compared with it. For example, a finish of 6.3 micrometres produced by grinding may well cost four times as much as rough machining. Many factors contribute towards production costs, and this information again can be only approximate.

Chapter 15
Screw threads and conventional representations

Screw threads

The most common application of the helix is in a screw thread which follows the path of the helix. Screw threads may be either left or right hand and these are shown pictorially in Fig. 15.1. Notice the slope of the thread and the position of the index finger on each hand. The left hand thread is used for special applications and the right hand thread is the one normally used on nuts and bolts. The thread illustrated has a vee-section.

Left hand thread

Right hand thread

Fig. 15.1

The following terms are associated with screw threads:

The *thread pitch* is the distance between corresponding points on adjacent threads. Measurements must be taken parallel to the thread axis.

The *major diameter* or *outside diameter* is the diameter over the crests of the thread, measured at right angles to the thread axis.

The *crest* is the most prominent part of the thread, internal or external.

The *root* lies at the bottom of the groove between two adjacent threads.

The *flank* of the thread is the straight side of the thread between the crest and root.

The *minor diameter, root diameter* or *core diameter* is the smallest diameter of the thread measured at right angles to the thread axis.

The *effective diameter* is measured at right angles to the thread axis and is the diameter on which the width of the spaces is equal to the width of the threads.

The *lead* of a thread is the axial movement of the screw in one revolution.

The terms are illustrated in Fig. 15.2.

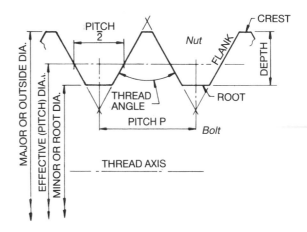

Fig. 15.2   External form to illustrate thread terms

ISO metric threads

Fig. 15.3 shows the ISO metric thread form for a nut (internal) and for a bolt (external). In the case of the nut, the root is rounded in practice. For the mating bolt, the crest of the thread may be rounded within the maximum outline, as shown, and the root radiused to the given dimension. Both male and female threads are subject to manufacturing tolerances and for complete information, reference should be made to BS 3643.

BS 3643 defines two series of diameters with graded pitches for general use in nuts, bolts and screwed fittings, one series with coarse and the other with fine pitches. The extract given in Table 15.1 from the standard gives thread sizes from 1.6 to 24 mm diameter. Note that first, second and third choices of basic diameters are quoted, to limit the number of sizes within each range.

On a drawing, a thread will be designated by the letter M followed by the size of the nominal diameter and the pitch required, e.g. M10 × 1

If a thread is dimensioned without reference to the pitch, e.g. M16, then it is assumed that the coarse-series thread is required.

Table 15.1

1	2	3	4	5	6	7	8	9	10	11	12	13	14	15	16
Basic major diameters			Coarse series with graded pitches	Fine series with constant pitches											
Choice															
1st	2nd	3rd		6	4	3	2	1.5	1.25	1	0.75	0.5	0.35	0.25	0.2
1.6	–	–	0.35	–	–	–	–	–	–	–	–	–	–	–	0.2
–	1.8	–	0.35	–	–	–	–	–	–	–	–	–	–	–	0.2
2	–	–	0.4	–	–	–	–	–	–	–	–	–	–	0.25	–
–	2.2	–	0.45	–	–	–	–	–	–	–	–	–	–	0.25	–
2.5	–	–	0.45	–	–	–	–	–	–	–	–	–	0.35	–	–
3	–	–	0.5	–	–	–	–	–	–	–	–	–	0.35	–	–
–	3.5	–	0.6	–	–	–	–	–	–	–	–	–	0.35	–	–
4	–	–	0.7	–	–	–	–	–	–	–	–	0.5	–	–	–
–	4.5	–	0.75	–	–	–	–	–	–	–	–	0.5	–	–	–
5	–	–	0.8	–	–	–	–	–	–	–	–	0.5	–	–	–
–	–	5.5	–	–	–	–	–	–	–	–	–	0.5	–	–	–
6	–	–	1	–	–	–	–	–	–	–	0.75	–	–	–	–
–	–	7	1	–	–	–	–	–	–	–	0.75	–	–	–	–
8	–	–	1.25	–	–	–	–	–	–	1	0.75	–	–	–	–
–	–	9	1.25	–	–	–	–	–	–	1	0.75	–	–	–	–
10	–	–	1.5	–	–	–	–	–	1.25	1	0.75	–	–	–	–
–	–	11	1.5	–	–	–	–	–	–	1	0.75	–	–	–	–
12	–	–	1.75	–	–	–	–	1.5	1.25	1	–	–	–	–	–
–	14	–	2	–	–	–	–	1.5	1.25*	1	–	–	–	–	–
–	–	15	–	–	–	–	–	1.5	–	1	–	–	–	–	–
16	–	–	2	–	–	–	–	1.5	–	1	–	–	–	–	–
–	–	17	–	–	–	–	–	1.5	–	1	–	–	–	–	–
–	18	–	2.5	–	–	–	2	1.5	–	1	–	–	–	–	–
20	–	–	2.5	–	–	–	2	1.5	–	1	–	–	–	–	–
–	22	–	2.5	–	–	–	2	1.5	–	1	–	–	–	–	–
24	–	–	3	–	–	–	2	1.5	–	1	–	–	–	–	–

Note. For preference, choose the diameters given in Column 1. If these are not suitable, choose from Column 2, or finally from Column 3.
*The pitch of 1.25 mm for 14 mm diameter is to be used only for sparking plugs.

Fig. 15.3 ISO metric thread
$H = 0.86603P$, $\frac{H}{4} = 0.21651P$, $\frac{3}{8}H = 0.32476P$
$\frac{5}{8}H = 0.54127P$, where P is the pitch of the thread

Unified threads

The Unified system of screw threads was introduced by the United Kingdom, Canada and the United States to provide a common standard thread for use by the three countries. The International Standards Organisation (ISO) recommends the system as an international system of screw threads in inch units, in parallel with a similar system in metric units. Both of these systems use a similar form of thread profile which is illustrated basically in Fig. 15.4.

Fig. 15.4 Unified screw thread
$H = 0.86603P$, $Rc = 0.108P$ and $Rr = 0.144P$ where P is the pitch of the thread

Unified threads are covered by BS 1580. Types in common use include the following;

a) UNC is a Unified coarse pitch thread with progressive pitch sizes (i.e. the pitch varies with the diameter).
b) UNF is a Unified fine pitch thread, also with progressive pitch sizes.
c) UN is a Unified thread with a constant pitch (e.g. an 8 UN thread has 8 threads to the inch regardless of the diameter).

Different classes of fit are obtainable by manufacture within alternative tolerance ranges and these are specified in BS 1580 Normally the same class of internal and external thread are used together.

Sellers or American thread (Fig. 15.5). This type was the American National thread in common use before the introduction of the Unified National thread, as it is described in the USA and Canada, or the Unified screw-thread in Great Britain.

Whitworth thread (Fig. 15.6). The general shape of the thread shown has been used in a standard BSW

Fig. 15.5 Sellers or American thread
$H = 0.866P$, $H' = 0.6495P$, $F = 0.1083P = H/8 = H/6$

Fig. 15.6 Basic Whitworth form
$H = 960491P$, $h = 0.640327P$, $r = 0.137329P$

thread, in fine form as the BSF thread, and as a pipe thread in the BSP thread.

The *British Standard Whitworth* thread was the first standardised British screw-thread.

The *British Standard Fine* thread is of Whitworth section but of finer pitch. The reduction in pitch increases the core diameter; also, small adjustments of the nut can easily be made.

The *British Standard Pipe* threads are used internally and externally on the walls of pipes and tubes. The thread pitch is relatively fine, so that the tube thickness is not unduly weakened.

Pipe threads are covered by BS 21, which was adopted as the basis for ISO 7/1 where the metric values were conversions of the inch values, to obtain interchangeability.

The basic forms of the British Standard taper and parallel pipe threads are based on that of the British Standard Whitworth thread.

The Whitworth thread form is shown in Fig. 15.6. The thread angle of 55° is measured in an axial plane section, also the vee-section is truncated, at top and bottom, by one sixth, with crest and root rounded by equal circular arcs. The theoretical thread depth is shown as $h = 0.640\,327\,P$ where P is the thread pitch. In the taper pipe thread (Fig. 15.7), a taper of 1 in 16 is used, measured on diameter.

British Association thread (Fig. 15.8). Generally used in sizes of less than $\frac{1}{4}$ inch on small mechanisms. This range of threads extends down to a thread size of 0.25 mm and is covered by BS 93.

Note. BS 93 is an 'obsolescent' standard. The standard is not recommended for the design of new equipment, but is retained to provide a standard for servicing equipment in use which is expected to have a long working life.

Fig. 15.7   Basic Whitworth form of taper pipe thread
$H = 0.960273P$, $h = 0.640327P$, $r = 0.137278P$

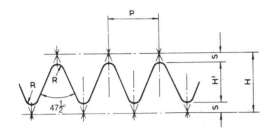

Fig. 15.8   British Association (BA) thread
$H = 1.1363P$, $H = 0.6P$ (aprox.), $R = 0.18P$, $S = 0.268P$

BS 4827 specifies the requirements for ISO miniature screw threads from 0.30 mm to 1.4 mm diameter with 60° form and are used in delicate instruments and watch making.

Threads for power transmission

Square thread (Fig. 15.9)

Used to transmit force and motion since it offers less resistance to motion than 'V' thread forms.

Widely used on lathes, this thread form is sometimes slightly modified by adding a small taper of about 5° to the sides as an aid to production.

Fig. 15.9

Acme thread (Fig. 15.10)

More easily produced than the square thread and often used in conjunction with split nuts for engagement purposes. It is applied in valve operating spindles.

Fig. 15.10   Acme thread
$C = 0.3707P$, $R = C$, $D = P/2 + 0.01$

Buttress thread (Fig. 15.11)

Used for transmitting power in one direction only. In its original form, the pressure face now sloping at 7° was perpendicular with the thread axis. A common application of the thread can be found in workshop vices.

Fig. 15.11   Buttress thread
$H = 0.8906P$, $H' = 0.5058P$, $S = 0.1395P$
$S' = 0.245P$. $F = 0.2754P$. $R = 0.1205P$

Draughting conventions associated with threads

Threads are so regularly used on engineering drawings that recognisable conventions are required to save draughting time. Fig. 15.12 shows the convention for a male thread. The thread runout along the shank of the stud is indicated by a line drawn at 30° to the thread axis. The minor diameter of the thread is shown by parallel lines and in the end elevation the projected circle is not continuous. The break in the inside circle distinguishes the end elevation of a male thread from a female thread. Line thicknesses are given for each part of the thread. The actual dimensions of the minor diameter for any particular thread size can be approxi-

Fig. 15.12

Fig. 15.13

mated at 80% of the major diameter for the purposes of conventional representation.

Fig. 15.13 shows the convention for a female thread applied to a blind tapped hole in a sectional view. Note that the minor diameter is drawn as a complete circle in the end elevation, the major diameter is broken and the different line thicknesses also help to distinguish the female from the male thread. The effective length of the thread is again shown by parallel lines and the runout by the taper at 30° to the thread axis. In the sectional elevation, the section lines drawn at 45°, continue through the major to the minor diameter. The included angle left by the tapping drill is 120°. Line thicknesses are indicated in the circles.

Note. The line at 30° indicating the runout was the practice in BS 308:1972. However, in 1984 the revised standard permitted the line to be omitted if there was no functional need. The tapered lines for incomplete threads are shown here for completeness of text.

A female thread through a collar is shown in Fig. 15.14. Note that the projection of the major and minor diameters drawn as hidden detail will be indicated by lines of short dashes 0.35 mm thick.

Fig. 15.14

A section through the collar is given in Fig. 15.15. The projection of the major diameter is drawn by a continuous 0.35 mm thick line and the cross hatching extends to the minor diameter.

A section through a sealing cap in Fig. 15.16 illustrates a typical application where an internal thread terminates in an undercut.

Screw threads are produced by cutting or rolling. A cut thread can be made by the use of a tap for female threads, or a die in the case of a male thread. Fig. 15.17 shows a male thread cut by a die and terminating in a runout. In this application the bar diameter has been drawn equal in size with the major diameter of the thread.

Fig. 15.15

Fig. 15.16

Fig. 15.17

Screwcutting may be undertaken on the lathe and the cutting tool is shaped to match the thread angle. Generally the thread terminates in an undercut and this feature is illustrated in Fig. 15.18. It is normal draughting practice to draw an undercut in line with the minor diameter of the thread. Too narrow an undercut or the demand for perfect or full threads up to shoulders, or to the bottom of blind holes, increases manufacturing costs.

Fig. 15.18

A rolled thread application is indicated in Fig. 15.19 The thread is formed by deformation and runs out to a bar diameter which is approximately equal to the effective diameter of the thread.

Fig. 15.19

Multiple threads

Generally, screws have single threads and unless it is designated otherwise, it is fair to assume that a thread will be single. This means that the thread is formed from one continuous helix. The lead of a thread is the distance moved by a mating nut in one complete revolution of the nut. In a single thread the lead is equal to the pitch.

When a two start thread is manufactured, there are two continuous helices and to accommodate the grooves, the lead is twice the thread pitch. Multiple threads are used where a quick axial movement is required with a minimum number of turns.

Now the standard drawing conventions do not differentiate between single and multiple threads. Details of the thread must be quoted with the drawing dimensions.

Fig. 15.20   Triple start right hand thread

(a)

(b)

Fig. 15.21

The application of thread conventions

Fig. 15.21(a) shows an elevation of a stud in a tapped hole. When a mating thread is screwed into a tapped hole, the section lines do not cover the assembled threads, also, the threads on the stud terminate at the start of the hole to signify that the stud is screwed to maximum depth. Adjacent parts in the sectional view are cross hatched in opposite directions. It is not normal practice to section nuts, bolts and washers.

Exercise—Draw a similar assembly using a M24 stud. The stud is screwed for 30 mm into the tapped hole and the thickness of the component with the clearance hole is 25 mm. Use Table 16.1 for other dimensions and assume proportions of unimportant detail.

Fig. 5.21(b) shows part of a component which is held by a set-bolt in a tapped hole. Only part of the shank of the bolt is threaded and this must terminate above the join line between the components for correct assembly.

Exercise—Draw a similar arrangement using a M20 set-bolt which engages with the tapped thread for a length of 25 mm. Component thickness is 22 mm. Assume other dimensions.

Note—The thickness of a nut is slightly greater than the thickness of a bolthead. For approximate constructions of a bolthead, use a thickness of $0.7D$, where D is the shank diameter.

Tapping drill

The diameter of a tapping drill is equal to the minor diameter of a thread to be tapped for drawing purposes. The function of a tapping drill is to remove the bulk of the material for a female thread, leaving the tap to cut only the thread form between the major and minor diameters.

Clearance drill

A clearance drill has a diameter slightly greater than the major diameter of a male thread and its function is to provide a clearance hole to permit the free passage of a bolt through a component.

Chapter 16
Nuts, bolts, screws and washers

ISO metric precision hexagon bolts, screws and nuts are covered by BS 3692 and ISO 272. The standard includes washer faced hexagon head bolts and full bearing head bolts. In both cases there is a small radius under the bolthead which would not normally be shown on drawings, due to its size, but is included here for completeness of the text. With an M36 bolt, the radius is only 1.7 mm. Bolts may be chamferred at 45° at the end of the shank, or radiused. The rounded end has a radius of approximately one and one quarter times the shank diameter and can also be used if required to draw the rolled thread end. The washer face under the head is also very thin and for a M36 bolt is only 0.5 mm.

Fig. 16.1(a) shows the bolt proportions and Table 16.1 the dimensions for bolts in common use. Dimensions of suitable nuts are also given and illustrated in Fig. 16.1(b).

Included in Table 16.1 and shown in Fig. 16.1(c) are typical washers to suit the above bolts and nuts and these are covered by BS 4320. Standard washers are available in two different thicknesses, in steel or brass, and are normally plain, but may be chamferred.

Table 16.1 gives dimensions of commonly used bolts, nuts and washers so that these can be used easily on assembly drawings. For some dimensions maximum and minimum values appear in the standards and we have taken an average figure rounded up to the nearest 0.5 mm and this will be found satisfactory for normal drawing purposes. Reference should be made to the relevant standards quoted for exact dimensions if required in design and manufacture.

Drawing nuts and bolts

It is often necessary to draw nuts and bolts and a quick easy method is required to produce a satisfactory result.

Table 16.1

Nominal size thread diameter D	Thread pitch	Minor diameter of thread	Width across corners A/C	Width across flats A/F	Diameter of washer face Df	Height of bolt head H	Thickness of normal nut T	Thickness of thin nut t	Washer inside diameter	Washer outside diameter	Washer thickness Form A	Washer thickness Form B
M1.6	0.35	1.1	3.5	3.0		1.0	1.25		1.7	4.0	0.3	
M2	0.4	1.4	4.5	4.0		1.5	1.5		2.2	5.0	0.3	
M2.5	0.45	1.9	5.5	5.0		1.75	2.0		2.7	6.5	0.5	
M3	0.5	2.3	6.0	5.5	5.0	2.0	2.25		3.2	7.0	0.5	
M4	0.7	3.0	8.0	7.0	6.5	2.75	3.0		4.3	9.0	0.8	
M5	0.8	3.9	9.0	8.0	7.5	3.5	4.0		5.3	10.0	1.0	
M6	1.0	4.7	11.5	10.0	9.0	4.0	5.0		6.4	12.5	1.6	0.8
M8	1.25	6.4	15.0	13.0	12.0	5.5	6.5	5.0	8.4	17	1.6	1.0
M10	1.5	8.1	19.5	17.0	16.0	7.0	8.0	6.0	10.5	21	2.0	2.5
M12	1.75	9.7	21.5	19.0	18.0	8.0	10.0	7.0	13.0	24	2.5	1.6
M16	2.0	13.5	27.0	24.0	23.0	10.0	13.0	8.0	17.0	30	3.0	2.0
M20	2.5	16.7	34.0	30.0	29.0	13.0	16.0	9.0	21.0	37	3.0	2.0
M24	3.0	20.0	41.5	36.0	34.5	15.0	19.0	10.0	25.0	44	4.0	2.5
M30	3.5	25.5	52.0	46.0	44.5	19.0	24.0	12.0	31.0	56	4.0	2.5
M36	4.0	31.0	62.5	55.0	53.5	23.0	29.0	14.0	37.0	66	5.0	3.0

HEXAGON HEAD
WASHER FACED

LENGTH   H

FULL BEARING
HEAD

ROLLED THREAD END

(a) Hexagon head precision bolts

(b) Precision nuts and thin nuts

(c) Bright washers

FORM A   FORM B

30°

CHAMFER
OPTIONAL

Fig. 16.1 Proportions of bolts, nuts and washers. A/C means across corners. A/F means across flats.

Nuts and bolts are not normally drawn on detail drawings unless they are of a special type. They are shown on assembly drawings and, provided they are standard stock sizes, are called up in parts lists and schedules. A description of the head, the thread and the length being generally sufficient. Templates are available for drawing nuts and bolts and can be recommended for their time saving advantages.

It is conventional drawing practice to show, as first choice, nuts and bolts in the across corners position if a single view only is illustrated since this is instantly recognisable.

Approximate construction for nuts and bolts (Figs 16.2 and 16.3)

Stage 1

1   Draw a circle in the plan position, 2D in diameter,

where D is equal to the thread size. In this example let us assume that the thread size is M20.

2   Draw a hexagon inside the 40 mm diameter circle and inside the hexagon draw another circle tangential to the hexagon on the six sides. This circle is the projection of the chamfer which can be seen on the front elevation.

3   The nut thickness is 0.8D. Project the four corners of the hexagon to the front elevation.

4   Project three corners of the hexagon in the end elevation and note, that the width of the end elevation is given by dimension W.

5   Line in the projected diameter of the chamfer circle and the base in the front elevation.

6   As an approximation, draw a radius to show the chamfer on the front elevation. The radius equal to the thread size D.

7   Add the female convention to the plan view.

Stage 2

1   The projection of the curve on the chamferred faces of the hexagon that lie at an angle would produce ellipses in the front elevation. In their place we usually show small circular arcs, their radii can be

Fig. 16.2 Stage 1

Fig. 16.3 Stage 2

found by trial, but are approximately 0.25D.

2 The end elevation of the nut has square corners and the projection of the corner which coincides with the centre line terminates at the bottom of the chamfer curve.

3 Complete the view by drawing circular arcs on the two chamferred faces. Find by trial, the radius of an arc which will touch the top of the nut and the projection lines from the corner in the front elevation.

Reference to Fig. 16.1a and b will show that the constructions in Fig. 16.2 and Fig. 16.3 can be used for the bolthead and locknut where proportions for thickness can be approximated to 0.7D and 0.5D.

For exact dimensions however, please refer to Table 16.1.

Socket head screws manufactured to BS 4168 and BS 3643 class 6g

It is often required to draw these screws and although the head type and the length are generally quoted in parts lists it is necessary to know the proportions of the head. Dimensions follow for each of the most commonly used screws.

Before specifying screws it is advisable to consult a manufacturers list for availability. In the interest of standardisation and economy, designers are urged to use stock lengths wherever possible and standard lengths of screws include the following; 3, 4, 5, 6, 8, 10, 12, 16, 20, 25, 30, 35, 40, 45, 50, 55, 60, 65, 70, 75, 80, 90, 100, 110, 120, 130, 140, 150, 160, 170, 180, 190, and 200 mm. If lengths over 200 mm are required, then increments of 20 mm are the preferred ISO lengths. It should be understood that not all diameters of screw are available in the above lengths. For example, the

range of lengths for an M3 screw lies between 5 and 35 mm, for an M10 screw between 12 and 100 mm for one particular type of head. The same range will also not cover different types of head, hence the necessity to check stock lists.

ISO metric socket cap screws Dimensions in Table 16.2

These screws are distinguished by square knurling on the heads. Generally, the lengths of standard screws increase in increments of 5 mm and 10 mm, but the exact range should be checked from the manufacturers catalogue.

Fig. 16.4 ISO metric hexagon socket shoulder screws
Dimensions in Table 16.3

Table 16.2

Nominal size D	M3	M4	M5	M6	M8	M10	M12	M16	M20
Head diameter A	5.5	7	8.5	10	13	16	18	24	30
Head depth H	3	4	5	6	8	10	12	16	20
Key engagement K	1.3	2	2.7	3.3	4.3	5.5	6.6	8.8	10.7
Socket size J	2.5	3	4	5	6	8	10	14	17

ISO metric hexagon socket shoulder screws
Dimensions in Table 16.3

Table 16.3

Nominal shoulder diameter B	6	8	10	12	16
Head diameter A	10	13	16	18	24
Head height H	4.5	5.5	7	8	10
Socket size J	3	4	5	6	8
Screw thread diameter D	M5	M6	M8	M10	M12
Nominal thread length Lt	9.75	11.25	13.25	16.4	18.4
Key engagement K	2.45	3.3	4.15	4.92	6.62

Fig. 16.5   ISO metric hexagon socket button head screws
Dimensions in Table 16.4

ISO metric hexagon socket button head screws
Dimensions in Table 16.4

Table 16.4

Nominal size D	M3	M4	M5	M6	M8	M10	M12
Head diameter A	5.5	7.5	9.5	10.5	14	18	21
Head depth H	1.6	2.1	2.7	3.2	4.3	5.3	6.4
Socket size J	2	2.5	3	4	5	6	8
Key engagement K	1.04	1.3	1.56	2.08	2.6	3.12	4.16
Fillet radius F-minimum	0.1	0.2	0.2	0.25	0.4	0.4	0.6
da-maximum	3.6	4.7	5.7	6.8	9.2	11.2	14.2
S	0.38	0.38	0.5	0.8	0.8	0.8	0.8

Fig. 16.6

ISO metric socket countersunk head screws
Dimensions in Table 16.5

The basic requirement for countersunk head screws is that the head should fit into a countersunk hole with as great a degree of flushness as possible, Fig. 16.7 and Table 16.5 give dimensions. To achieve this it is necessary for both the head of the screw and the countersunk hole to be controlled within prescribed limits. The maximum or design size of the head is controlled by a theoretical diameter to a sharp corner and the minimum head angle of 90°. The minimum head size is controlled by a minimum head diameter, the maximum head angle of 92° and a flushness tolerance. The edge of the head may be flat, as shown in Fig. 16.8, or rounded but not sharp edged (see below).

Fig. 16.7

Fig. 16.8   ISO metric hexagon socket set screws
Dimensions in Table 16.6

Table 16.5

Nominal size D	M3	M4	M5	M6	M8	M10	M12	M16	M20
Head diameter									
A – maximum	6.72	8.96	11.2	13.44	17.92	22.4	26.88	33.6	40.32
B – minimum	5.82	7.78	9.78	11.73	15.73	19.67	23.67	29.67	35.61
Head depth H	1.86	2.48	3.1	3.72	4.96	6.2	7.44	8.8	10.16
Socket size J	2	2.5	3	4	5	6	8	10	12
Key engagement K	1.05	1.49	1.86	2.16	2.85	3.60	4.35	4.89	5.45
Fillet radius									
F – minimum	0.1	0.2	0.2	0.25	0.4	0.4	0.6	0.6	0.8
da – maximum	3.4	4.5	5.5	6.6	9	11	14	18	22

ISO metric hexagon socket set screws
Dimensions in Table 16.6

These screws are available with a variety of pointed ends. In all cases the overall length includes the chamfer at the socket end and the point.

Table 16.6

Nominal size D	M3	M4	M5	M6	M8	M10	M12	M16	M20
Socket size J	1.5	2	2.5	3	4	5	6	8	10
Key engagement K	1.2	1.6	2	2.4	3.2	4	4.8	6.4	8
Dog point diameter P	2	2.5	3.5	4.5	6	7	9	12	15
Dog point length Q	0.88	1.12	1.38	1.62	2.12	2.62	3.12	4.15	5.15
Cup point and 'W' point diameter C	1.4	2	2.5	3	5	6	8	10	14

Note: the cone point angle shown as $Y°$ is generally 118° for short screws and 90° for longer lengths.

Fig. 16.9 (a) Cone point (b) Dog point (c) W point (d) Cup point

Fig. 16.10 Typical GKN socket screws and wrench

Machine screws

Head shapes for machine screws have been rationalized in BS 4183. For the purpose of this British Standard, the generic term 'screws' applies to products which are threaded up to the head or having an unthreaded portion of the shank.

The length of the thread is defined as the distance from the end of the screw, and this includes any chamfer, radius or cone point, to the leading face of the nut which has been screwed as far as possible onto the screw by hand. Note on the illustrations which follow that in the case of the countersunk head types of screw, the length of the screw includes the countersunk part of the head. For pan and cheese head screws, the screw length does not include the head.

The Standard should be consulted for manufacturing dimensional tolerances, also concentricity tolerances for the heads of the screws.

The illustrations which follow show each of the screws and tables are also given showing the dimensions of regularly used sizes.

The sizes quoted in the tables are for screws manufactured in steel. Standard screws are also available in brass but generally the range is not quite so extensive.

For all of the machine screws illustrated here, the countersunk head types have an included angle of 90°.

'Posidriv' machine screws, countersunk and raised countersunk head
Dimensions in Table 16.7

Fig. 16.11 Countersunk head

Fig. 16.12 Raised countersunk head

Table 16.7

Diameter d	Dia of head D	Depth of head H	Driver no.	Length L										
M2	4.40	1.20	1	4	5	6	8	10	12					
M2.5	5.50	1.50	1	5	6	8	10	12	16	20	25			
M3	6.30	1.65	1	5	6	8	10	12	16	20	25			
M3.5	7.35	1.93	2	6	8	10	12	16	20	25	30			
M4	8.40	2.20	2	6	8	10	12	16	20	25	30			
M5	10.00	2.50	2	6	8	10	12	16	20	25	30	35	40	45	50
M6	12.00	3.00	3	10	12	16	20	25	30	35	40	45	50	55	60
M8	16.00	4.00	4	12	16	20	25	30	40	50	60			
M10	20.00	5.00	4	16	20	25	30	40	50	60				

'Posidriv' machine screws, pan head
Dimensions in Table 16.8

Fig. 16.13

Table 16.8

Diameter d	Dia of head D	Depth of head H	Driver no.	Length L											
M2	4.00	1.60	1	4	5	6	8	10	12						
M2.5	5.00	1.75	1	5	6	8	10	12	16	20	25				
M3	6.00	2.10	1	5	6	8	10	12	16	20	25				
M3.5	7.00	2.45	2	6	8	10	12	16	20	25	30				
M4	8.00	2.80	2	5	6	8	10	12	16	20	25	30			
M5	10.00	3.50	2	6	8	10	12	16	20	25	30	35	40	45	50
M6	12.00	4.20	3	10	12	16	20	25	30	35	40	45	50	55	60
M8	16.00	5.60	4	16	20	25	30	40	50						
M10	20.00	7.00	4	20	25	30	40								

Slotted machine screws, countersunk and raised countersunk head
Dimensions in Table 16.9

Fig. 16.14 Countersunk head

Fig. 16.15 Raised countersunk head

Table 16.9

Diameter d	Dia of head D	Depth of head H	Length L															
M2	4.40	1.20	5	6	8	10	12											
M2.5	5.50	1.50	5	6	8	10	12	16	20	25								
M3	6.30	1.65	5	6	8	10	12	16	20	25	30	35	40	45	50			
M3.5	7.35	1.93	5	6	8	10	12	16	20	25	30	35	40	45	50	60		
M4	8.40	2.20	5	6	8	10	12	16	20	25	30	35	40	45	50	60	70	
M5	10.00	2.50	6	8	10	12	16	20	25	30	35	40	45	50	60	70	80	
M6	12.00	3.00	8	10	12	16	20	25	30	35	40	45	50	60	70	80	90	
M8	16.00	4.00	10	12	16	20	25	30	35	40	45	50	60	70	80	90		
M10	20.00	5.00	16	20	25	30	35	40	45	50	55	60	70	80	90			

Slotted machine screws, pan head
Dimensions in Table 16.10

Fig. 16.16

Table 16.10

Diameter d	Dia of head D	Depth of head H	Length L													
M2	4.00	1.20	4	5	6	8	10	12								
M2.5	5.00	1.50	5	6	8	10	12	16	20	25						
M3	6.00	1.80	5	6	8	10	12	16	20	25	30	35	40	45	50	
M3.5	7.00	2.10	6	8	10	12	16	20	25	30	35	40	45	50		
M4	8.00	2.40	5	6	8	10	12	16	20	25	30	35	40	45	50	
M5	10.00	3.00	6	8	10	12	16	20	25	30	35	40	45	50	55	60	70	80
M6	12.00	3.60	8	10	12	16	20	25	30	35	40	45	50	55	60	70	80
M8	16.00	4.80	10	12	16	20	25	30	40	50	60	70	80	90		
M10	20.00	6.00	16	20	25	30	40	50	60	70						

Slotted machine screws, cheese head
Dimensions in Table 16.11

Fig. 16.17

Table 16.11

Diameter d	Dia of head D	Depth of head H	Length L														
M2	3.80	1.30	3	4	5	6	8	10	12	16	20	25					
M2.5	4.50	1.60	5	6	8	10	12	16	20	25	30						
M3	5.50	2.00	4	5	6	8	10	12	16	20	25	30	35	40	45	50	
M3.5	6.00	2.40	5	6	8	10	12	16	20	25	30	35	40	45	50	60	70
M4	7.00	2.60	5	6	8	10	12	16	20	25	30	35	40	45	50	60	70
M5	8.50	3.30	6	8	10	12	16	20	25	30	35	40	45	50	60	70	80	90
M6	10.00	3.90	8	10	12	16	20	25	30	35	40	45	50	60	70	80	90
M8	13.00	5.00	10	12	16	20	25	30	35	40	45	50	60	70	80	90	
M10	16.00	6.00	16	20	25	30	35	40	45	50	60	70	80	90			

Machine screw nuts

A range of machine screw nuts is covered by BS 4183 and these nuts are manufactured in two different patterns, square and hexagon. The table shows typical nuts for use with the screws previously described.

Fig. 16.19 Hexagon nut

Fig. 16.18 Square nut

Table 16.12 Machine screw nuts, pressed type, square and hexagonal

Nominal size of nut d	Width across flats s	Width across corners e		Thickness
		Square	Hexagon	m
M2	4.0	5.7	4.6	1.2
M2.5	5.0	7.1	5.8	1.6
M3	5.5	7.8	6.4	1.6
M3.5	6.0	8.5	6.9	2.0
M4	7.0	9.9	8.1	2.0
M5	8.0	11.3	9.2	2.5
M6	10.0	14.1	11.5	3.0
M8	13.0	18.4	15.0	4.0
M10	17.0	24.0	19.6	5.0

Wing nuts

Fig. 16.20 shows the dimensions of preferred sizes of wing nuts manufactured in brass or malleable iron by a hot stamping or casting process.

An alternative wing nut is available in brass or malleable iron and manufactured by cold forging.

Fig. 16.20 (Sizes are in Table 16.13)

Table 16.13

Thread size	A	B	C	D	E	F	G	H	J
M3	9	6.5	7	13.5	22	19	3.5	2.5	1.5
M4 and M5	10	8	9	15	25.5	19	4	2.5	1.5
M6	13	9.5	11	18	30	19	5	2.5	1.5
M8	16	12	13	23	38	19	6.5	3	2.5
M10	17.5	14	14	25.5	44.5	19	7	5	3
M12	19	16	15	28.5	51	25.5	8	5	3
M16	25.5	20.5	19	36.5	63.5	32	10	6.5	5

Locking and retaining devices

The function of a locking device is to prevent loosening or disengagement of mating components which may be operating under varying conditions of stress, temperature and vibration. The effectiveness of the locking device may be vital to safety.

One of the simplest locking devices is a locknut and these are generally thin plain nuts which are tightened against ordinary plain nuts or against components into which male threaded items are assembled. To ensure efficient locking, the bearing surfaces of the nut and component must bed together evenly and the correct degree of tightness obtained by applying the designed torque loading. The locknut should not be over-tightened as this may result in the stripping of the nut threads or overstressing of the male component. In cases where rotation can occur, the plain nut must be held stationary whilst the locknut is tightened.

Fig. 16.21 Bolts, nut and locknut

Slotted nuts and castle nuts

One method of preventing nuts from coming loose is to drill the bolt and use a pin through the assembly. Suitable nuts are shown in Fig. 16.22. Slotted nuts are available for sizes M4 to M39 and have six slots. Castle nuts are also available with six slots between sizes M12 to M39 and eight slots between sizes M42 and M68. For convenience in drawing both types of nuts, the total

(a) Slotted nut   **(b) Castle nut**

Fig. 16.22

thickness can be approximated to the thread diameter plus 2 millimetres. The dimensions for the hexagons can be taken from Table 16.1.

Slotted nuts are reusable but difficult to apply where access is limited.

Simmond's locknut

This type of locknut incorporates a collar manufactured from nylon or fibre and the collar is slightly smaller in diameter than the internal thread diameter. The section in Fig. 16.23 shows the collar in black. On assembly, the stud or bolt forces its way through the resilient collar which provides a frictional lock. The locknut is a little thicker than a conventional nut.

Fig. 16.23

Spring washers

This type of washer is produced as a single or a double coil spring. The cross section is rectangular. Generally this type of washer dispenses with the simple plain washer although a plain washer can be used at the same time with assemblies where the component is manufactured from relatively soft light alloys. The free height of double coil washers before compression is normally about five times the thickness of the steel section.

Fig. 16.24

Table 16.14 Double coil rectangular section spring washers to BS 4464

Nominal thread diameter	Max. inside diameter I/D	Max. outside diameter O/D	Thickness S	Free height H
M2	2.4	4.4	0.50	2.50
M3	3.6	6.2	0.80	4.00
M4	4.6	8.0	0.80	4.00
M5	5.6	9.8	0.90	4.50
M6	6.6	12.9	1.00	5.00
M8	8.6	15.1	1.20	6.00
M10	10.8	18.2	1.20	6.00
M12	12.8	20.2	1.60	8.00
M16	17.0	27.4	2.00	10.00
M20	21.5	31.9	2.00	10.00
M24	26.0	39.4	3.25	16.25
M30	33.0	49.5	3.25	16.25
M36	40.0	60.5	3.25	16.25

Lockwasher with external teeth
(a) Flat form   (b) Dished type

Dished-type washer with toothed periphery

Lockwasher with
internal teeth

Single-coil
washer

Fig. 16.25   Types of locking washer

Shakeproof washers

This type of washer is generally made from spring steel and serrations are formed on either the internal or external diameters. These serrations then bite into the pressure faces between the nut and the component when the nut is assembled. Some slight disfiguration of the component may result on assembly but this is of little significance except where anti-corrosion treatment of the component surface has previously been carried out. Some screws are pre-assembled with conical lockwashers which are free to rotate but do not come off.

Toothed lockwashers combat vibration and are especially suited to rough parts or surfaces.

Wire locking

Non-corrodible steel and brass wire, of the appropriate gauge, are normally used for wire locking. Generally, a hole is provided for this purpose in the component to be locked and the wire is passed through and twisted. The lay of the wire between the anchorage and the component must always be such as to resist any tendency of the locked part or parts to become loose.

Fig. 16.26 shows the plan view of a pressurized cylinder and the cover is held down by four bolts which are wire locked. The operation is performed with a single strand of wire. The wire is passed in sequence through the holes in the bolts and the ends are twisted together to tension the wire loop. Note, that in order to become loose, the bolts must turn in an anticlockwise direction but this will have the effect of increasing the

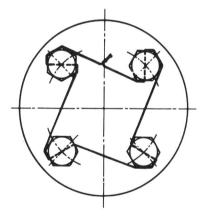

Fig. 16.26

tension in the wire loop. The locking wire should only be used once.

Tab washers

Tab washers are thin metal washers designed with two or more tabs which project from the external diameter. On assembly, a tab is bent against the component or sometimes into a hole in the component. Other tabs are

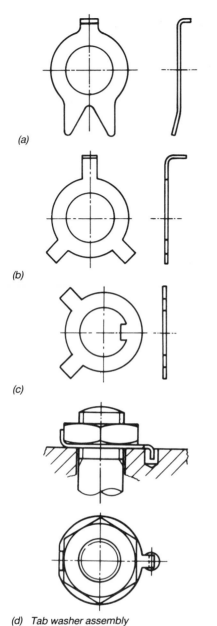

(a)

(b)

(c)

(d)　*Tab washer assembly*

Fig. 16.27

Fig. 16.28　Typical locking plate for a hexagonal nut

(a)　Flat type

(b)　Bent type

(c)　Flat-wing type

Fig. 16.29　Locking terminals

then bent up against the correctly tightened nut. Another pattern has a tab projecting from the inside diameter and this is intended to fit into a slot machined in the bolt, whilst the external tabs are again bent against the flat sides of the nut. The deformation of the tab washer is such that it is intended to be used only once.

Three different types of tab washer are shown in Fig. 16.27 together with, a typical assembly.

Locking plates

Locking plates are manufactured usually from mild steel and fit over hexagonal nuts after these have been tightened on assembly. The locking plate is then

secured on the component by a separate screw which may itself be fitted with a shakeproof or spring type of washer.

Locking plates may be used repeatedly, provided they remain a good fit, around the hexagon of the nut or the bolthead. Locking plates may be cranked, as in Fig. 16.28 or flat.

Fig. 16.29 shows a selection of locking terminals where a 'Shakeproof' washer and a soldering lug are combined into one unit, thus saving assembly time. The locking teeth anchor the terminal to the base, to prevent shifting of the terminal in handling, while the twisted teeth produce a multiple bite which penetrates an oxidised or painted surface to ensure good conductivity. All three types of locking terminal are generally made from phosphor bronze with a hot-tinned finish.

Taper pins and parallel pins

Taper pins, with a taper of 1 in 50, and parallel pins are used on both solid and tubular sections to secure, for example, levers to torque shafts and control rods to fork ends. Some taper pins are bifurcated, or split, and the legs can be opened out for locking. Plain taper pins and parallel pins may also be locked by peening. To prevent slackness, these pins are assembled in accurate reamed holes. Undue force should not be used during the peening process or the security of the fittings may be impaired if the pin is bent.

Fig. 16.30(a) shows part of a lever which is fixed to a hollow operating shaft by a bifurcated taper pin. On assembly, a hole is drilled which is slightly smaller than the diameter at the small end of the taper pin and this is enlarged by a taper pin reamer so that the small end of the taper pin, when pushed through the assembly, is flush with the surface. The pin is then driven into position. If the pin is of the bifurcated type, then the legs are spread to form an included angle of about 60°. Fig. 16.30(b) shows the same operating lever assembled, but using a parallel pin, which has been peened over after ensuring that the component is adequately supported.

Fig. 16.31 shows the general shape of a taper pin. Parallel sides are substituted for tapered sides in parallel pins.

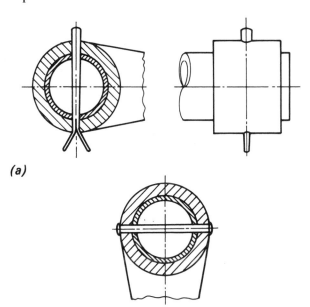

(a)

(b)

Fig. 16.30

Split cotter pins

Ferrous and non ferrous split cotter pins are covered by BS 1574. The designating size of a split cotter pin is the size of the hole for which it is intended to fit. When reference is made to a split cotter pin in a parts list, this

Fig. 16.31

Fig. 16.32 Proportions of split cotter pins to BS 1574

Table 16.15 Split cotter pin dimensions

Nominal diameter of pin (hole) diameter	Shank diameter d	Outside diameter of eye c	Length of eye b	Length extended prong a
1.0	0.9	1.8	3.0	1.6
1.2	1.0	2.0	3.0	2.5
1.6	1.4	2.8	3.2	2.5
2.0	1.8	3.6	4.0	2.5
2.5	2.3	4.6	5.0	2.5
3.2	2.9	5.8	6.4	3.2
4.0	3.7	7.4	8.0	4.0
5.0	4.6	9.2	10.0	4.0
6.3	5.9	11.8	12.6	4.0
8.0	7.5	15.0	16.0	4.0
10.0	9.5	19.0	20.0	6.3
13.0	12.4	24.8	26.0	6.3

nominal dimension is followed by the length required. The closed legs of the shank of the pin form a circular cross section. The legs should be straight and parallel throughout their nominal length. Fig. 16.32 shows alternative pins in detail.

Exact dimensions are given in BS 1574 but for drawing purposes the following extracts will be useful where the maximum allowable sizes for dimensions a, b, c and d are quoted. Many standard lengths are obtainable for each pin size and as a rough guide between 5 and 25 times the shank diameter.

Locking by adhesives

Small components found in, for example, instruments and switches may be locked by the application of Shellac, Araldite, Loctite, or similar materials. Shellac and Loctite are usually applied to the threads of nuts, bolts, screws and studs and the components are assembled while still wet. The parts should be free from grease to achieve maximum strength. Araldite is applied to the outside of the nut face and the protruding screw thread, after tightening. Araldite is an adhesive which hardens, after mixing, within a specified time period.

Peening

This operation prevents re-use of the screw or bolt but locking can be carried out by peening over about $1\frac{1}{2}$ threads. This practice can be used in the case of screwed pivots and a simple example is often found in a pair of scissors. In the case of nuts and bolts, peening is carried down to the nut to prevent it from slackening.

Countersunk screws may be locked by peening metal from the surroundings into the screw slot. This practice is sometimes adopted when the thread is inaccessible.

Thread-cutting screws

Barber and Colman Ltd are the manufacturers of 'Shakeproof' thread-cutting screws and washers.

'Shakeproof' thread-cutting screws made from carbon steel are subjected to a special heat-treatment which provides a highly carburised surface with a toughened resilient core. The additional strength provided enables higher tightening torques to be used, and will often permit the use of a smaller-size thread-cutting screw than would normally be specified for a machine screw. Thread-cutting screws actually cut their own mating thread; in any thickness of material a perfect thread-fit results in greatly increased holding power, extra vibration-resistance, and a faster assembly. The hard, keen cutting edge produces a clean-cut thread, from which the screw can be removed, if desired, without damage to screw or the cut thread. The most suitable drill sizes for use with these screws are generally larger than standard tapping-drill sizes, but this

Type 1   Type 23

Fig. 16.33

Fig. 16.34   Type 25 thread-cutting screw

apparent loss of thread engagement is more than offset by the perfect thread-fit obtained.

Both the screws shown in Fig. 16.33 are interchangeable with standard machine screws. Type 1 is recommended for use in steel and non-ferrous sheet and plate, and they are manufactured with a wide shank slot and are eminently suitable for paint-clearing applications, as they completely eliminate the need for expensive pre-production tapping of painted assemblies. Type 23 screws incorporate a special wide cutting slot with an acute cutting angle for fast, easy thread-cutting action and ample swarf clearance. These screws are specially designed for application into soft metals or plastics where a standard thread form is required.

The Type 25 thread-cutting screw has a specially spaced thread form which is designed for fast efficient fastening into plastics and sheet-metal applications.

Fig. 16.35 illustrates a 'Teks' self-drilling screw which, with a true drilling action, embodies three basic operations in one device. It (1) prepares its own hole, (2) either cuts or forms a mating thread, and (3) makes a complete fastening in a single operation. These screws

Fig. 16.35   'Teks' self-drilling screw

consist of an actual drill point to which a threaded screw-fastener has been added. Several different head styles are available. During the drilling stage, Teks must be supported rigidly from the head. Some bench-mounted, automatically fed screwdrivers provide a holding means which retracts as the screw is finally driven home. Other drivers connect with the fastener only through the bit or socket. A good-fitting Phillips or Pozidriv bit will normally drive several thousand of these screws, and a hex socket, for hex-head designs, will drive even more.

For long screws or applications requiring absolutely guaranteed driving stability, a special chuck is available which holds the screw with three fingers and retracts upon contacting the work surface. These screws are suitable for fastening sheet steel of 16 gauge, or thicker, within 5 seconds maximum while using a power tool.

Fig. 16.36 shows alternative head styles available for thread-cutting screws.

The examples and dimensions of nuts, bolts, screws and washers given here are intended especially to be of use to students engaged on design projects. There are however literally hundreds of industrial fastening systems available, associated with automobile, construction, electronics and aerospace developments. Manufacturers' catalogues are freely available to provide technical specifications and necessary details for designers. One further advantage of CAD systems is that such information can be used to build a library of useful data and drawings, which are invaluable, where contract drawings use a repetition of similar parts.

(a) Slotted round

(b) Slotted pan

(c) Slotted cheese

(d) Slotted fillister

(e) Pozidriv flange

(f) Posidriv pan

Fig. 16.36

Chapter 17
Keys and keyways

A key, Fig. 17.1, is usually made from steel and is inserted between the joint of two parts to prevent relative movement; it is also inserted between a shaft and a hub in an axial direction, to prevent relative rotation. A keyway, Figs. 17.2, 17.3 and 17.4, is a recess in a shaft or hub to receive a key, and these recesses are commonly cut on key-seating machines or by broaching, milling, planing, shaping and slotting. The proportions of cross-sections of keys vary with the shaft size, and reference should be made to BS 4325 for the exact dimensions. The length of the key controls the area of the cross-section subject to shear, and will need to be calculated from a knowledge of the forces being transmitted or, in the case of feather keys, the additional information of the length of axial movement required.

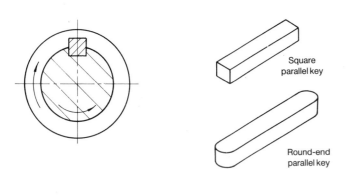

Square parallel key

Round-end parallel key

Gib-head key

Fig. 17.1

Fig. 17.2   Edge-milled keyway

Fig. 17.3   End-milled keyway

Fig. 17.4   Keyway in hub

Sunk keys

Examples of sunk keys are shown in Fig. 17.5, where the key is sunk into the shaft for half its thickness. This measurement is taken at the side of the key, and not along the centre line through the shaft axis. Fig. 17.5 shows useful proportions used for assembly drawings.

Square and rectangular keys may be made with a taper of 1 in 100 along the length of the key; Fig. 17.6 shows such an application. Note that, when dimensioning the mating hub, the dimension into the keyway is taken across the maximum bore diameter.

A *gib head* may be added to a key to facilitate removal, and its proportions and position when assembled are given in Fig. 17.7.

A *feather key* is attached to either the shaft or the hub, and permits relative axial movement while at the same time enabling a twisting moment to be transmitted between shaft and hub or vice versa. Both pairs of opposite faces of the key are parallel.

A *double-headed feather key* is shown in Fig. 17.8 and allows a relatively large degree of sliding motion between shaft and hub. The key is inserted into the bore of the hub, and the assembly is then fed on to the shaft, thus locking the key in position.

Rectangular key

D = SHAFT DIAMETER

$$W = \frac{D}{4}$$

$$T = \frac{D}{6}$$

Square key

Fig. 17.5

TAPER 1:100
ON THIS FACE

Fig. 17.6

Gib-head key

TAPER 1:100
ON TOP SURFACE

45°

Fig. 17.7

Fig. 17.8   Double-headed feather key

A *peg feather key* is shown in Fig. 17.9, where a peg attached to the key is located in a hole through the hub.

Fig. 17.10 illustrates a feather key which is screwed in position in the shaft keyway by two countersunk screws.

Hub

Shaft

Fig. 17.9   Peg feather key

Fig. 17.10   Feather key

Woodruff keys

A Woodruff key, Fig. 17.11, is a segment of a circular disc and fits into a circular recess in the shaft which is machined by a Woodruff keyway cutter. The shaft may be parallel or tapered, Figs 17.12 and 17.13 showing the method of dimensioning shafts for Woodruff keys where the depth of the recess from the outside of the shaft is given, in addition to the diameter of the recess. A Woodruff key has the advantage that it will turn itself in its circular recess to accommodate any taper in the mating hub on assembly; for this reason it cannot be used as a feather key, since it would jam. Woodruff keys are commonly used in machine tools and, for example, between the flywheel and the crankshaft of a small internal-combustion engine where the drive depends largely on the fit between the shaft and the conically bored flywheel hub. The deep recess for a

Fig. 17.11   Woodruff key

Fig. 17.12   Dimensions required for a Woodruff key in a parallel shaft

Fig. 17.13   Dimensions required for a Woodruff key in a tapered shaft

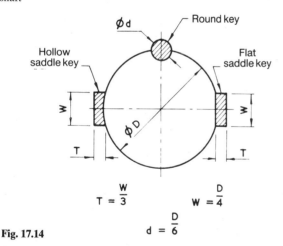

Fig. 17.14

$$T = \frac{W}{3} \qquad W = \frac{D}{4}$$
$$d = \frac{D}{6}$$

Woodruff key weakens the shaft, but there is little tendency for the key to turn over when in use.

Where lighter loads are transmitted and the cost of cutting a keyway is not justified, round keys and flat or hollow saddle keys as shown in Fig. 17.14 can be used.

Saddle keys are essentially for light duty only, overloading tending to make them rock and work loose on the shaft. Both flat and hollow saddle keys may have a

taper of 1 in 100 on the face in contact with the hub. The round key may either be tapered or, on assembly, the end of the shaft and hub may be tapped after drilling and a special threaded key be screwed in to secure the components.

Dimensioning keyways (parallel keys)

The method of dimensioning a parallel shaft is shown in Fig. 17.15, and a parallel hub in Fig. 17.16. Note that in each case it is essential to show the dimension to the bottom of the keyway measured across the diameter of the shaft and the bore of the hub. This practice cannot be used where either the shaft or hub is tapered, and Fig. 17.17 shows the method of dimensioning a keyway for a square or rectangular parallel key in a tapered shaft, where the keyway depth is shown from the outside edge of the shaft and is measured vertically into the bottom of the slot. Fig. 17.18 shows a tapered hub with a parallel keyway where the dimension to the bottom of the slot is taken across the major diameter. A parallel hub utilising a tapered key is also dimensioned across the major diameter, as indicated in Fig. 17.19.

Fig. 17.15   Keyway in parallel shaft

Fig. 17.16   Keyway in parallel hub

Fig. 17.17   Keyway for square or rectangular parallel key in tapered shaft

Fig. 17.18   Tapered hub with parallel keyway

Fig. 17.19   Parallel hub with tapered keyway

Chapter 18
Worked examples in machine drawing

Examination questions generally relate to single parts or assemblies of detailed components, and test the students' ability to draw sectional and outside views. British Standards refer to 'views' but other terms have been traditionally used in technical drawing. The front or side view of a house is quite likely to be known as an elevation and a 'birds eye view' as a plan. These expressions are freely used.

The examples which follow are of examination standard and a student draughtsman would be expected to produce a reasonably complete solution for each problem in about two to three hours on a drawing board.

Before commencing, try and estimate the areas covered by the views so that they can be presented with reasonably equal spaces horizontally and vertically on the drawing sheet. Include a border about 15 mm width, add a title block and parts list if necessary. Always print between parallel guide lines. Note that attention to small details will gradually enable you to improve the quality of your draughtsmanship.

In an industrial situation, before commencing a drawing, the draughtsman will make a mental picture of how to orient the component, or arrangement, so that the maximum amount of information can be indicated with the minimum number of views necessary to produce a clear unambiguous solution. However, this is easier said than done in the case of the student, and especially where the drawing is being made on CAD

Fig. 18.1

**ENLARGED SECTION
THROUGH TEE SLOT**

Fig. 18.2   Note: all unspecified radii 6

equipment since the size of the screen often means that part of the drawing is temporarily out of sight. This is part of the learning experience.

Copy the following solutions and try to appreciate the reasons for the position of every line which contributes to the finished drawing.

1 Bushed bearing bracket

Front and end views of a bushed bearing bracket are shown in Fig. 18.1. Copy the given front view and project from it a sectional end view and a sectional plan view taken from cutting planes A–A and B–B. Draw

your solution in first angle projection. Note that the question is presented in third angle projection.

2 Drill table

Fig. 18.2 shows details of a table for a drilling machine. Draw half full size the following views:

a) A front view taken as a section along the cutting plane A–A.
b) The given plan view with hidden detail.
c) An end view projected to the left of the front view with hidden detail included.

Draw your solution in first angle projection and add the title and projection symbol in a suitable title block.

Fig. 18.3

① CLAMP BASE

② OPERATING ARM

③ CAM PLATE

④ STOPSCREW

4–Ø8 HOLES

BRAZE
Ø10 ROD

BRAZE

M10 × 1.5
× 15 DEEP

(5) PLUNGER

5 TURNS OF Ø3
WIRE GROUND FLAT
AT EACH END

FREE LENGTH

(6) SPRING

(7) ARM PIVOT

(8) CAM PIVOT

NOTE:- ON ASSEMBLY, FIT
A SUITABLE WASHER OVER
EACH PIVOT AND DRILL FOR
2 mm Ø SPLIT PIN

2 M10 / 1.5
LOCKNUTS
AND WASHER

KEY TO ASSEMBLY

Fig. 18.3 *(continued)*

3 Cam operated clamp

Details of a cam operated clamp are shown in Fig. 18.3. together with a key showing the position of the various components in the assembly. Draw the following views in first angle projection:

a) A front view taken as a section through the vertical centre line of the clamp and generally as outlined in the key.
b) An end view projected on the right-hand side of the front view
c) A plan view drawn beneath the front view

Add to your solution a title, scale, projection symbol, parts list and reference balloons.

(1) VALVE BODY

Fig. 18.4

4 Plug cock

Details of a plug cock are shown in Fig. 18.4. Draw the following two views of the assembled plug cock:

a) A view in half section showing the outside view on the left of the vertical centre line and a section on the right hand side, the position of the valve body being similar to that shown on the given detail.

b) An outside end view

Add a title and a parts list. Hidden details are not required in either view. Include on your assembly suitable nuts and bolts and washers where applicable. Between the valve body and cover include a joint ring of thickness 2 mm. The gland should be positioned entering 10 mm into the valve cover. First or third angle projected views will be acceptable.

Fig. 18.4 *(continued)*

5 Air engine

The component parts of an oscillating air engine are detailed in Fig. 18.5. Draw in first angle projection and, at twice full size, the following views;

a) A front view taken as a section through the engine cylinder and flywheel.

b) A plan view in projection with the front view and drawn below the front view

c) An end view on the right hand side and in projection with the front view.

Hidden detail is not required in any view. Add a parts list to your drawing and reference balloons to identify each of the components in the assembly.

Fig. 18.5

CONNECTING PIPES 2 OFF 20 SWG
SWEAT IN PORTS

LIGHT STEEL COMPRESSION
SPRING ID 5.0

MED. KNURL

Fig. 18.5 *(continued)*

6 Toolbox

The details in Fig. 18.6 show parts of a toolbox which can be made as a sheet-metal-work exercise.

A development is given of the end plates (two required) and these should be drilled at the corners before bending, to avoid possible cracking taking place. Note, from the photograph, that the ends have been pressed with grooves to provide additional stiffening and this can be done if facilities are available.

The bottom and sides are fabricated in one piece and drillings are shown for pop rivets. Drill the end plates using the sides as templates to take account of deformation which may occur while bending the metal plate.

Details are shown of the hinges and some clearance must be provided for their satisfactory operation. Before drilling the sides and lids for the hinges using the dimensions given, use the hinges as manufactured to check the hole centres for the rivets. One lid must overlap the other to close the box, so some adjustment

PART 1
END PLATES
2 REQ'D 1mm MS

DEVELOPMENT
OF END PLATE

DRILL Ø2 HOLES
AT EACH CORNER
BEFORE BENDING

PART 3
BOX LID
2 REQ'D 1mm MS

AFTER DRILLING, FOLD
ALONG BEND LINE TO
THIS POSITION

PART 2
BOTTOM AND SIDE PANELS
1 REQ'D 1mm M.S.

ALL HOLES Ø3

BEND LINE

Ø3 HOLES
FIXINGS TO
ENDPLATES

196 INSIDE

PICTORIAL VIEW (SCALE 1:4)
SHOWING FOLDED SIDES TO
INTERNAL RADIUS OF 2mm
ALSO BOTTOM HOLES FOR
RIVETTING TO ENDPLATES

NOTE:- ALL CORNERS MAY BE
CUT BACK 10mm TO FIT FLUSH
WITH ENDPLATES

Fig. 18.6

DRILL 2–Ø3 HOLES
EACH END ON ASSEMBLY

PART 4
TOP SUPPORT
1 REQ'D 1mm M.S.
ASSEMBLE UNDER END PLATES
BY POP RIVETTING

3 HOLES
Ø3

BLANK FOR HINGE
8 REQ'D

ROLL HINGE BLANK
TO SUIT M.S. ROD
Ø4 × 102 LONG

PEEN OVER PIN
AT EACH END

PART 5
ASSEMBLED HINGE
4 REQ'D 1mm M.S.

2 HOLES Ø3

PART 6
HASP
1 REQ'D 1mm M.S.

PEEN OVER PIN
AT EACH END

ROLL HINGE BLANK
TO SUIT M.S. PIN
Ø4 × 26 LONG

PART 7
CATCH
2mm M.S.
1 REQ'D

Ø5 M.S. ROD   R5

PART 8
HANDLE
2 REQ'D

2 HOLES Ø3   R7

PART 9
HANDLE CLIPS
4 REQ'D 1mm M.S.

FOLD UP FOUR
SIDES. WELD OR
BRAZE CORNERS

PART 10
TOOL TRAY
1 REQ'D 1mm M.S.

BEND LINES

TOOL TRAY
DEVELOPMENT

Ø3 HOLE
EACH END

FOLD UP BOTH
ENDS. WELD OR
BRAZE CORNERS

PART 11
TRAY SUPPORT BRACKET
2 REQ'D 1mm M.S.

DEVELOPMENT

Fig. 18.6 (*continued*)

here may be required. On the toolbox illustrated, pop rivets were used on the sides and roundhead rivets on the lid. Roundhead rivets were also used to assemble the handle clips. Check also the drill sizes for the rivets used as alternatives are available.

The top support member was pop rivetted to the end plates after the catch had been welded on the underside. The slot in each lid slips over this catch to allow a padlock to secure the toolbox.

A tooltray may also be made if required and details are provided. The tray supports also offer added stability and were pop rivetted on assembly.

Remove all sharp corners and edges before assembly of each part. The completed toolbox can be stove enamelled, painted or sprayed according to personal choice.

As an additional exercise, copy the given details and produce an assembly drawing complete with parts list. Also, produce patterns to cut the components from sheet metal economically. This example provides useful practice in reading and applying engineering drawings.

Solution shown in Fig. 18.12 (and see Fig. 18.13).

Solution notes

Fig. 18.7   Bushed bearing bracket

Note that in the end view there is an example of a thin web which is not cross hatched. The example shows three examples of counterbored holes used to contain heads of fixing screws.

A liner bush with an interference fit is also indicated, and since the bush is relatively thin compared with the main casting, its cross hatching lines are drawn closer together.

Fig. 18.8   Drill table

The table is clamped to the drilling machine main vertical pillar using two bolts. The slot permits tightening.

Cross hatching is omitted on the left side of the front view.

Fig. 18.7

TITLE

DRILL TABLE

FIRST ANGLE PROJECTION

DIMENSIONS IN MILLIMETRES | SCALE:

DRAWN BY: | APPROVED BY:

Fig. 18.8

Fig. 18.9   Cam operated clamp

This example shows a typical assembly and includes a parts list.

The balloons containing the part numbers are equally spaced whenever possible. The leader lines to the components terminate in a dot. The leader line is also directed towards the centre of the balloon and touches the circumference.

It is considered good policy to position the parts list and commence writing in such a way that parts can be added on to the end of the list in the event of future modifications. The list can be drawn at the top or bottom of the drawing sheet.

Fig. 18.10   Plug cock assembly

This example illustrates a typical industrial valve. The plug turns through 90° between the on and off positions. Spanner flats are provided and indicated by the diagonal lines. Gland packing (Part 10) is supplied in rings. These rings are contained by the body cover and on assembly are fed over the plug spindle. The gland is tightened and the compressive force squeezes the packing to provide a seal, sufficient to prevent leaks, but enabling the spindle to be turned. The joint ring (Part 9) is too thin to be cross hatched and is shown filled in.

This is also an example of a symmetrical part where the half section gives an outside view and a sectional view to indicate the internal details.

Fig. 18.11   Air engine assembly, illustrated below

The engine operates through compressed air entering the cylinder via one of the connecting pipes shown as item 15. The other pipe serves to exhaust the cylinder after the power stroke. The cylinder oscillates in an arc and a hole through the cylinder wall lines up with the inlet and exhaust pipes at each 180° of rotation of the flywheel.

The spindles (Parts 4 and 18), grub screw (Part 5) and the pin (Part 11) would not normally be sectioned. A part section is illustrated at the bottom of the connecting rod in order to show its assembly with the crank pin (Part 11). The BS convention is shown for the spring (Part 16). The BS convention is also shown for cross knurling on the nut (Part 17).

Note. If the solutions to examples 3, 4, 5 and 6 had been required to be drawn in third angle projection, then the views would have been arranged as shown in Fig. 18.13.

10	M10 WASHER	3
9	M10 LOCKNUT	2
8	CAM PIVOT	1
7	ARM PIVOT	1
6	SPRING	1
5	PLUNGER	1
4	STOPSCREW	1
3	CAM PLATE	1
2	OPERATING ARM	1
1	CLAMP BASE	1
PART Nº	DESCRIPTION	Nº OFF

TITLE **ASSEMBLY OF CAM OPERATED CLAMP**

FIRST ANGLE PROJECTION

DIMENSIONS IN MILLIMETRES   SCALE:

DRAWN BY: APPROVED BY:

Fig. 18.9

ITEM No.	DESCRIPTION	No. OFF
1	VALVE BODY	1
2	PLUG	1
3	VALVE BODY COVER	1
4	GLAND	1
5	COVER STUD	4
6	GLAND STUD	2
7	WASHER	6
8	NUT	6
9	JOINT RING	1
10	GLAND PACKING	4

HALF SECTION A - A

TITLE   PLUG COCK ASSEMBLY

PROJECTION	FIRST ANGLE	PROJECTION SYMBOL	
DIMENSIONS IN MILLIMETRES		ORIGINAL SCALE	1 : 1
ALL UNSPECIFIED RADII 3MM		DRAWN BY	

Fig. 18.10

ITEM No.	DESCRIPTION	No. OFF	MATERIAL
1	BASE	1	BDMS
2	UPRIGHT	1	BDMS
3	FLYWHEEL	1	BDMS
4	CRANKSHAFT	1	BDMS
5	GRUB SCREW	1	BDMS
6	BACK PLATE	1	BRASS
7	CYLINDER	1	BRASS
8	HEAD	1	BRASS
9	PISTON	1	BDMS
10	CONNECTING ROD	1	BDMS
11	CRANK PIN	1	BDMS
12	CRANK WEB	1	BDMS
13	BEARING HOUSING	1	BDMS
14	BUSH	2	BRASS BRONZE
15	PIPES	2	COPPER
16	SPRING	1	STEEL
17	KNURLED NUT	1	BRASS
18	CYLINDER PIVOT	1	BDMS

TITLE	AIR ENGINE ASSEMBLY	
PROJECTION	PROJECTION SYMBOL	
DIMENSIONS IN MILLIMETRES	ORIGINAL SCALE	2 : 1
ALL UNSPECIFIED RADII 3MM	DRAWN BY	

Fig. 18.11

Fig. 18.12

PART No.	DESCRIPTION	No. OFF
11	TRAY SUPPORT BRACKET	2
10	TOOL TRAY	1
9	HANDLE CLIPS	4
8	HANDLE	2
7	HASP	1
6	CATCH	1
5	HINGES	4
4	TOP SUPPORT	1
3	LID	2
2	BOTTOM & SIDE PANELS	1
1	END PLATE	2

TOOL BOX ASSEMBLY

SCALE - HALF FULL SIZE	FIRST ANGLE PROJECTION

A – A

Fig. 8.13

Chapter 19
Limits and fits

To ensure that an assembly will function correctly, its component parts must fit together in a predictable manner. Now, in practice, no component can be manufactured to an exact size, and one of the problems facing the designer is to decide the upper and lower limits of size which are acceptable for each of the dimensions used to define shape and form and which will ensure satisfactory operation in service. For example, a dimension of 10 ± 0.02 means that a part will be acceptable if manufactured anywhere between the limits of size of 9.98 mm and 10.02 mm. The present system of manufacture of interchangeable parts was brought about by the advent of and the needs of mass production, and has the following advantages.

1 Instead of 'fitting' components together, which requires some adjustment of size and a high degree of skill, they can be 'assembled'.
2 An assembly can be serviced by replacing defective parts with components manufactured to within the same range of dimensions.
3 Parts can be produced in large quantities, in some cases with less demand on the skill of the operator. Invariably this requires the use of special-purpose machines, tools, jigs, fixtures, and gauges: but the final cost of each component will be far less than if made separately by a skilled craftsman.

It should be noted, however, that full interchangeability is not always necessary in practice; neither is it always feasible, especially when the dimensions are required to be controlled very closely in size. Many units used in the construction of motor vehicles are assembled after an elaborate inspection process has sorted the components into different groups according to size. Suppose, for example, that it was required to maintain the clearance between a piston and a cylinder to within 0.012 mm. To maintain full interchangeability would require both the piston and the cylinder bores to be finished to a tolerance of 0.006 mm, which would be difficult to maintain and also uneconomic to produce. In practice it is possible to manufacture both bores and pistons to within a tolerance of 0.06 mm and then divide them into groups for assembly; this involves the gauging of each component.

A designer should ensure that the drawing conveys clear instructions regarding the upper and lower limits of size for each dimension, and Figs 19.1 to 19.4 show typical methods in common use.

The method shown in Fig. 19.1 is perhaps the clearest

Fig. 19.1

way of expressing limits of size on a drawing, since the upper and lower limits are quoted, and the machine operator is not involved in mental arithmetic. The dimensions are quoted in logical form, with the upper limit above the lower limit and both to the same number of decimal places.

As an alternative to the method above, the basic size may be quoted and the tolerance limits added as in Fig. 19.2. It is not necessary to express the nominal dimension to the same number of decimal places as the limits.

Fig. 19.2

Fits can be taken directly from those tabulated in BS 4500, 'ISO limits and fits', and, in order to indicate the grade of fit, the following alternative methods of dimensioning a hole may be used:

$$90 \text{ H7} \left(\frac{90.035}{90.000} \right) \text{ (first choice)}$$

$$\text{or} \quad 90 \text{ H7} \quad \text{or} \quad 90 \text{ H7} \left(\begin{array}{c} +0.035 \\ 0 \end{array} \right)$$

Similarly, a shaft may be dimensioned as follows:

$$90 \text{ g6} \begin{pmatrix} 89.988 \\ 89.966 \end{pmatrix} \text{ (first choice)}$$

$$\text{or} \quad 90 \text{ g6} \quad \text{or} \quad 90 \text{ g6} \begin{pmatrix} -0.012 \\ -0.034 \end{pmatrix}$$

In cases where a large amount of repetition is involved, information can be given in tabulated form, and a typical component drawing is shown in Fig. 19.3.

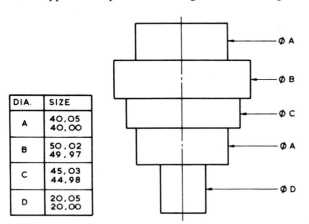

DIA.	SIZE
A	40,05 40,00
B	50,02 49,97
C	45,03 44,98
D	20,05 20,00

Fig. 19.3

In many cases, tolerances need be only of a general nature, and cover a wide range of dimensions. A box with a standard note is added to the drawing, and the typical examples in Fig. 19.4 are self explanatory.

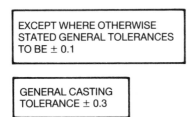

EXCEPT WHERE OTHERWISE STATED GENERAL TOLERANCES TO BE ± 0.1

GENERAL CASTING TOLERANCE ± 0.3

Fig. 19.4

Engineering fits between two mating parts can be divided into three types:

1 a *clearance fit* (Fig. 19.5), in which the shaft is always smaller than the hole into which it fits;
2 an *interference fit* (Fig. 19.6), in which the shaft is always bigger than the hole into which it fits;
3 a *transition fit* (Fig. 19.7), in which the shaft may be either bigger or smaller than the hole into which it fits—it will therefore be possible to get interference or clearance fits in one group of assemblies.

It will be appreciated that, as the degree of accuracy required for each dimension increases, the cost of production to maintain this accuracy increases at a sharper rate.

Fig. 19.8 shows the approximate relationship between cost and tolerance. For all applications, the manufacturing tolerance should be the largest possible which permits satisfactory operation.

Fig. 19.5   Clearance fits—allowance always positive

Fig. 19.6   Interference fits—allowance always negative

Fig. 19.7   Transition fit—allowance may be positive or negative

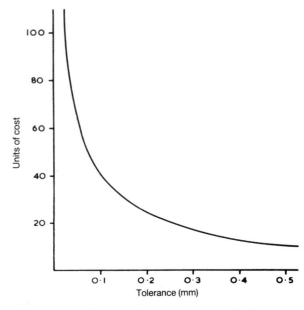

Fig. 19.8   Approximate relationship between production cost and manufacturing tolerance

Elements of interchangeable systems (Fig. 19.9)

Fig. 19.9 Elements of interchangeable systems

Nominal size is the size by which a component is referred to as a matter of convenience, i.e. 25 mm, 50 mm, 60 mm thread.

Actual size is the measured size.

Basic size is the size in relation to which all limits of size are fixed, and will be the same for both the male and female parts of the fit.

Limits of size These are the maximum and minimum permissible sizes acceptable for a specific dimension.

Tolerance This is the total permissible variation in the size of a dimension, and is the difference between the upper and lower acceptable dimensions.

Allowance concerns mating parts, and is the difference between the high limit of size of the shaft and the low limit of size of its mating hole. An allowance may be positive or negative.

Grade This is an indication of the tolerance magnitude: the lower the grade, the finer will be the tolerance.

Deviation This is the difference between the maximum, minimum, or actual size of a shaft or hole and the basic size.

Maximum metal condition (MMC) This is the maximum limit of an external feature; for example, a shaft manufactured to its high limit would contain the maximum amount of metal. It is also the minimum limit on an internal feature; for example, a component which has had a hole bored in it to its lower limit of size would have had the minimum of metal removed and remain in its maximum metal condition.

Unilateral and bilateral limits

Fig. 19.10 shows an example of unilateral limits, where the maximum and minimum limits of size are disposed on the same side of the basic size. This system is preferred since the basic size is used for the GO limit gauge; changes in the magnitude of the tolerance affect only the size of the other gauge dimension, the NOT GO gauge size.

Fig. 19.11 shows an example of bilateral limits, where the limits are disposed above and below the basic size.

Fig. 19.10 Unilateral limits

Fig. 19.11 Bilateral limits

Bases of fits

The two bases of a system of limits and fits are

a) the hole basis,
b) the shaft basis.

Hole basis (Fig. 19.12) In this system, the basic diameter of the hole is constant while the shaft size varies according to the type of fit. This system leads to greater economy of production, as a single drill or reamer size can be used to produce a variety of fits by merely altering the shaft limits. The shaft can be accurately produced to size by turning and grinding. Generally it is usual to recommend hole-base fits, except where temperature may have a detrimental effect on large sizes.

Fig. 19.12 Hole-basis fits: C – clearance T – transition I – interference

Shaft basis (Fig. 19.13) Here the hole size is varied to produce the required class of fit with a basic-size shaft. A series of drills and reamers is required for this system, therefore it tends to be costly. It may, however, be necessary to use it where different fits are required along a long shaft. The BSI data sheet 4500A gives a selection of ISO fits on the hole basis, and data sheet 4500B gives a selection of shaft-basis fits extracted from BS 45000, the current standard on limits and fits.

The ISO system contained in BS 4500 gives an extensive selection of hole and shaft tolerances to cover a wide range of applications. It has been found, however, that in the manufacture of many standard engineering components a limited selection of tolerances is adequate. These are provided on the data sheets referred to above. Obviously, by using only a selected range of fits, economic advantages are obtained from the reduced tooling and gauging facilities involved.

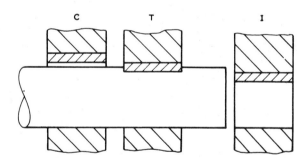

Fig. 19.13 Shaft-basis fits: C – clearance T – transition I – interference

Selected ISO fits—hole basis (extracted from BS 4500)

The ISO system provides a great many hole and shaft tolerances so as to cater for a very wide range of conditions. However, experience shows that the majority of fit conditions required for normal engineering products can be provided by a quite limited selection of tolerances. The following selected hole and shaft tolerances have been found to be commonly applied:

selected hole tolerances: H7 H8 H9 H11;
selected shaft tolerances: c11 d10 e9 f7 g6 h6 k6 n6 p6 s6.

Table 19.1 shows a range of fits derived from these selected hole and shaft tolerances. As will be seen, it covers fits from loose clearance to heavy interference, and it may therefore be found to be suitable for most normal requirements. Many users may in fact find that their needs are met by a further selection within this selected range.

It should be noted, however, that this table is offered only as an example of how a restricted selection of fits can be made. It is clearly impossible to recommend selections of fits which are appropriate to all sections of industry, but it must be emphasised that a user who decides upon a selected range will always enjoy the economic advantages this conveys. Once he has installed the necessary tooling and gauging facilities, he can combine his selected hole and shaft tolerances in different ways without any additional investment in tools and equipment.

For example, if it is assumed that the range of fits shown in the table has been adopted but that, for a particular application the fit H8-f7 is appropriate but provides rather too much variation, the hole tolerance H7 could equally well be associated with the shaft f7 and may provide exactly what is required without necessitating any additional tooling.

For most general applications, it is usual to recommend hole-basis fits, as, except in the realm of very large sizes where the effects of temperature play a large part, it is usually considered easier to manufacture and measure the male member of a fit, and it is thus desirable to be able to allocate the larger part of the tolerance available to the hole and adjust the shaft to suit.

In some circumstances, however, it may in fact be preferable to employ a shaft basis. For example, in the case of driving shafts where a single shaft may have to accommodate a variety of accessories such as couplings, bearings, collars, etc., it is preferable to maintain a constant diameter for the permanent member, which is the shaft, and vary the bore of the accessories. For use in applications of this kind, a selection of shaft basis fits is provided in data sheet BS 4500B.

Note. Data Sheet 4500A (p. 147) refers to hole basis fits.

Data Sheet 4500B (p. 148) refers to shaft basis fits.

BRITISH STANDARD
SELECTED ISO FITS—HOLE BASIS

Extracted from BS 4500: 1969

Data Sheet 4500A
Issue 1: February 1970
Confirmed August 1985

Diagram to scale for 25 mm diameter

Clearance fits · Transition fits · Interference fits

Holes · Shafts

All tolerance values in 0.001 mm.

Nominal sizes Over	To	H11	c11	H9	d10	H9	e9	H8	f7	H7	g6	H7	h6	H7	k6	H7	n6	H7	p6	H7	s6	Nominal sizes Over	To
—	3	+60 / 0	−60 / −120	+25 / 0	−20 / −60	+25 / 0	−14 / −39	+14 / 0	−6 / −16	+10 / 0	−2 / −8	+10 / 0	0 / −6	+10 / 0	+6 / 0	+10 / 0	+10 / +4	+10 / 0	+12 / +6	+10 / 0	+20 / +14	—	3
3	6	+75 / 0	−70 / −145	+30 / 0	−30 / −78	+30 / 0	−20 / −50	+18 / 0	−10 / −22	+12 / 0	−4 / −12	+12 / 0	0 / −8	+12 / 0	+9 / +1	+12 / 0	+16 / +8	+12 / 0	+20 / +12	+12 / 0	+27 / +19	3	6
6	10	+90 / 0	−80 / −170	+36 / 0	−40 / −98	+36 / 0	−25 / −61	+22 / 0	−13 / −28	+15 / 0	−5 / −14	+15 / 0	0 / −9	+15 / 0	+10 / +1	+15 / 0	+19 / +10	+15 / 0	+24 / +15	+15 / 0	+32 / +23	6	10
10	18	+110 / 0	−95 / −205	+43 / 0	−50 / −120	+43 / 0	−32 / −75	+27 / 0	−16 / −34	+18 / 0	−6 / −17	+18 / 0	0 / −11	+18 / 0	+12 / +1	+18 / 0	+23 / +12	+18 / 0	+29 / +18	+18 / 0	+39 / +28	10	18
18	30	+130 / 0	−110 / −240	+52 / 0	−65 / −149	+52 / 0	−40 / −92	+33 / 0	−20 / −41	+21 / 0	−7 / −20	+21 / 0	0 / −13	+21 / 0	+15 / +2	+21 / 0	+28 / +15	+21 / 0	+35 / +22	+21 / 0	+48 / +35	18	30
30	40	+160 / 0	−120 / −280	+62 / 0	−80 / −180	+62 / 0	−50 / −112	+39 / 0	−25 / −50	+25 / 0	−9 / −25	+25 / 0	0 / −16	+25 / 0	+18 / +2	+25 / 0	+33 / +17	+25 / 0	+42 / +26	+25 / 0	+59 / +43	30	40
40	50	+160 / 0	−130 / −290	+62 / 0	−80 / −180	+62 / 0	−50 / −112	+39 / 0	−25 / −50	+25 / 0	−9 / −25	+25 / 0	0 / −16	+25 / 0	+18 / +2	+25 / 0	+33 / +17	+25 / 0	+42 / +26	+25 / 0	+59 / +43	40	50
50	65	+190 / 0	−140 / −330	+74 / 0	−100 / −220	+74 / 0	−60 / −134	+46 / 0	−30 / −60	+30 / 0	−10 / −29	+30 / 0	0 / −19	+30 / 0	+21 / +2	+30 / 0	+39 / +20	+30 / 0	+51 / +32	+30 / 0	+72 / +53	50	65
65	80	+190 / 0	−150 / −340	+74 / 0	−100 / −220	+74 / 0	−60 / −134	+46 / 0	−30 / −60	+30 / 0	−10 / −29	+30 / 0	0 / −19	+30 / 0	+21 / +2	+30 / 0	+39 / +20	+30 / 0	+51 / +32	+30 / 0	+78 / +59	65	80
80	100	+220 / 0	−170 / −390	+87 / 0	−120 / −260	+87 / 0	−72 / −159	+54 / 0	−36 / −71	+35 / 0	−12 / −34	+35 / 0	0 / −22	+35 / 0	+25 / +3	+35 / 0	+45 / +23	+35 / 0	+59 / +37	+35 / 0	+93 / +71	80	100
100	120	+220 / 0	−180 / −400	+87 / 0	−120 / −260	+87 / 0	−72 / −159	+54 / 0	−36 / −71	+35 / 0	−12 / −34	+35 / 0	0 / −22	+35 / 0	+25 / +3	+35 / 0	+45 / +23	+35 / 0	+59 / +37	+35 / 0	+101 / +79	100	120
120	140	+250 / 0	−200 / −450	+100 / 0	−145 / −305	+100 / 0	−85 / −185	+63 / 0	−43 / −83	+40 / 0	−14 / −39	+40 / 0	0 / −25	+40 / 0	+28 / +3	+40 / 0	+52 / +27	+40 / 0	+68 / +43	+40 / 0	+117 / +92	120	140
140	160	+250 / 0	−210 / −460	+100 / 0	−145 / −305	+100 / 0	−85 / −185	+63 / 0	−43 / −83	+40 / 0	−14 / −39	+40 / 0	0 / −25	+40 / 0	+28 / +3	+40 / 0	+52 / +27	+40 / 0	+68 / +43	+40 / 0	+125 / +100	140	160
160	180	+250 / 0	−230 / −480	+100 / 0	−145 / −305	+100 / 0	−85 / −185	+63 / 0	−43 / −83	+40 / 0	−14 / −39	+40 / 0	0 / −25	+40 / 0	+28 / +3	+40 / 0	+52 / +27	+40 / 0	+68 / +43	+40 / 0	+133 / +108	160	180
180	200	+290 / 0	−240 / −530	+115 / 0	−170 / −355	+115 / 0	−100 / −215	+72 / 0	−50 / −96	+46 / 0	−15 / −44	+46 / 0	0 / −29	+46 / 0	+33 / +4	+46 / 0	+60 / +31	+46 / 0	+79 / +50	+46 / 0	+151 / +122	180	200
200	225	+290 / 0	−260 / −550	+115 / 0	−170 / −355	+115 / 0	−100 / −215	+72 / 0	−50 / −96	+46 / 0	−15 / −44	+46 / 0	0 / −29	+46 / 0	+33 / +4	+46 / 0	+60 / +31	+46 / 0	+79 / +50	+46 / 0	+159 / +130	200	225
225	250	+290 / 0	−280 / −570	+115 / 0	−170 / −355	+115 / 0	−100 / −215	+72 / 0	−50 / −96	+46 / 0	−15 / −44	+46 / 0	0 / −29	+46 / 0	+33 / +4	+46 / 0	+60 / +31	+46 / 0	+79 / +50	+46 / 0	+169 / +140	225	250
250	280	+320 / 0	−300 / −620	+130 / 0	−190 / −400	+130 / 0	−110 / −240	+81 / 0	−56 / −108	+52 / 0	−17 / −49	+52 / 0	0 / −32	+52 / 0	+36 / +4	+52 / 0	+66 / +34	+52 / 0	+88 / +56	+52 / 0	+190 / +158	250	280
280	315	+320 / 0	−330 / −650	+130 / 0	−190 / −400	+130 / 0	−110 / −240	+81 / 0	−56 / −108	+52 / 0	−17 / −49	+52 / 0	0 / −32	+52 / 0	+36 / +4	+52 / 0	+66 / +34	+52 / 0	+88 / +56	+52 / 0	+202 / +170	280	315
315	355	+360 / 0	−360 / −720	+140 / 0	−210 / −440	+140 / 0	−125 / −265	+89 / 0	−62 / −119	+57 / 0	−18 / −54	+57 / 0	0 / −36	+57 / 0	+40 / +4	+57 / 0	+73 / +37	+57 / 0	+98 / +62	+57 / 0	+226 / +190	315	355
355	400	+360 / 0	−400 / −760	+140 / 0	−210 / −440	+140 / 0	−125 / −265	+89 / 0	−62 / −119	+57 / 0	−18 / −54	+57 / 0	0 / −36	+57 / 0	+40 / +4	+57 / 0	+73 / +37	+57 / 0	+98 / +62	+57 / 0	+244 / +208	355	400
400	450	+400 / 0	−440 / −840	+155 / 0	−230 / −480	+155 / 0	−135 / −290	+97 / 0	−68 / −131	+63 / 0	−20 / −60	+63 / 0	0 / −40	+63 / 0	+45 / +5	+63 / 0	+80 / +40	+63 / 0	+108 / +68	+63 / 0	+272 / +232	400	450
450	500	+400 / 0	−480 / −880	+155 / 0	−230 / −480	+155 / 0	−135 / −290	+97 / 0	−68 / −131	+63 / 0	−20 / −60	+63 / 0	0 / −40	+63 / 0	+45 / +5	+63 / 0	+80 / +40	+63 / 0	+108 / +68	+63 / 0	+292 / +252	450	500

BRITISH STANDARDS INSTITUTION, 2 Park Street, London, W1A 2BS

SBN: 580 05766 6

BRITISH STANDARD

SELECTED ISO FITS—SHAFT BASIS

Extracted from BS 4500 : 1969

Data Sheet 4500B — Issue 1. February 1970

BRITISH STANDARDS INSTITUTION, 2 Park Street, London, WIY 4AA — SBN: 580 05567 1

Gr 2 plus VAT (Gr 7 per set of ten copies)

Diagram to scale for 25 mm diameter — Holes / Shafts

Values in 0.001 mm. Each cell shows upper limit / lower limit of tolerance.

Nominal sizes Over (mm)	To (mm)	h11	C11	h9	D10	h9	E9	h7	F8	h6	G7	h6	H7	h6	K7	h6	N7	h6	P7	h6	S7
—	3	0 / -60	+120 / +60	0 / -25	+60 / +20	0 / -25	+39 / +14	0 / -10	+20 / +6	0 / -6	+12 / +2	0 / -6	+10 / 0	0 / -6	0 / -10	0 / -6	-4 / -14	0 / -6	-6 / -16	0 / -6	-14 / -24
3	6	0 / -75	+145 / +70	0 / -30	+78 / +30	0 / -30	+50 / +20	0 / -12	+28 / +10	0 / -8	+16 / +4	0 / -8	+12 / 0	0 / -8	+3 / -9	0 / -8	-4 / -16	0 / -8	-8 / -20	0 / -8	-15 / -27
6	10	0 / -90	+170 / +80	0 / -36	+98 / +40	0 / -36	+61 / +25	0 / -15	+35 / +13	0 / -9	+20 / +5	0 / -9	+15 / 0	0 / -9	+5 / -10	0 / -9	-4 / -19	0 / -9	-9 / -24	0 / -9	-17 / -32
10	18	0 / -110	+205 / +95	0 / -43	+120 / +50	0 / -43	+75 / +32	0 / -18	+43 / +16	0 / -11	+24 / +6	0 / -11	+18 / 0	0 / -11	+6 / -12	0 / -11	-5 / -23	0 / -11	-11 / -29	0 / -11	-21 / -39
18	30	0 / -130	+240 / +110	0 / -52	+149 / +65	0 / -52	+92 / +40	0 / -21	+53 / +20	0 / -13	+28 / +7	0 / -13	+21 / 0	0 / -13	+6 / -15	0 / -13	-7 / -28	0 / -13	-14 / -35	0 / -13	-27 / -48
30	40	0 / -160	+280 / +120	0 / -62	+180 / +80	0 / -62	+112 / +50	0 / -25	+64 / +25	0 / -16	+34 / +9	0 / -16	+25 / 0	0 / -16	+7 / -18	0 / -16	-8 / -33	0 / -16	-17 / -42	0 / -16	-34 / -59
40	50	0 / -160	+290 / +130	0 / -62	+180 / +80	0 / -62	+112 / +50	0 / -25	+64 / +25	0 / -16	+34 / +9	0 / -16	+25 / 0	0 / -16	+7 / -18	0 / -16	-8 / -33	0 / -16	-17 / -42	0 / -16	-34 / -59
50	65	0 / -190	+330 / +140	0 / -74	+220 / +100	0 / -74	+134 / +60	0 / -30	+76 / +30	0 / -19	+40 / +10	0 / -19	+30 / 0	0 / -19	+9 / -21	0 / -19	-9 / -39	0 / -19	-21 / -51	0 / -19	-42 / -72
65	80	0 / -190	+340 / +150	0 / -74	+220 / +100	0 / -74	+134 / +60	0 / -30	+76 / +30	0 / -19	+40 / +10	0 / -19	+30 / 0	0 / -19	+9 / -21	0 / -19	-9 / -39	0 / -19	-21 / -51	0 / -19	-48 / -78
80	100	0 / -220	+390 / +170	0 / -87	+260 / +120	0 / -87	+159 / +72	0 / -35	+90 / +36	0 / -22	+47 / +12	0 / -22	+35 / 0	0 / -22	+10 / -25	0 / -22	-10 / -45	0 / -22	-24 / -59	0 / -22	-58 / -93
100	120	0 / -220	+400 / +180	0 / -87	+260 / +120	0 / -87	+159 / +72	0 / -35	+90 / +36	0 / -22	+47 / +12	0 / -22	+35 / 0	0 / -22	+10 / -25	0 / -22	-10 / -45	0 / -22	-24 / -59	0 / -22	-66 / -101
120	140	0 / -250	+450 / +200	0 / -100	+305 / +145	0 / -100	+185 / +85	0 / -40	+106 / +43	0 / -25	+54 / +14	0 / -25	+40 / 0	0 / -25	+12 / -28	0 / -25	-12 / -52	0 / -25	-28 / -68	0 / -25	-77 / -117
140	160	0 / -250	+460 / +210	0 / -100	+305 / +145	0 / -100	+185 / +85	0 / -40	+106 / +43	0 / -25	+54 / +14	0 / -25	+40 / 0	0 / -25	+12 / -28	0 / -25	-12 / -52	0 / -25	-28 / -68	0 / -25	-85 / -125
160	180	0 / -250	+480 / +230	0 / -100	+305 / +145	0 / -100	+185 / +85	0 / -40	+106 / +43	0 / -25	+54 / +14	0 / -25	+40 / 0	0 / -25	+12 / -28	0 / -25	-12 / -52	0 / -25	-28 / -68	0 / -25	-93 / -133
180	200	0 / -290	+530 / +240	0 / -115	+355 / +170	0 / -115	+215 / +100	0 / -46	+122 / +50	0 / -29	+61 / +15	0 / -29	+46 / 0	0 / -29	+13 / -33	0 / -29	-14 / -60	0 / -29	-33 / -79	0 / -29	-105 / -151
200	225	0 / -290	+550 / +260	0 / -115	+355 / +170	0 / -115	+215 / +100	0 / -46	+122 / +50	0 / -29	+61 / +15	0 / -29	+46 / 0	0 / -29	+13 / -33	0 / -29	-14 / -60	0 / -29	-33 / -79	0 / -29	-113 / -159
225	250	0 / -290	+570 / +280	0 / -115	+355 / +170	0 / -115	+215 / +100	0 / -46	+122 / +50	0 / -29	+61 / +15	0 / -29	+46 / 0	0 / -29	+13 / -33	0 / -29	-14 / -60	0 / -29	-33 / -79	0 / -29	-123 / -169
250	280	0 / -320	+620 / +300	0 / -130	+400 / +190	0 / -130	+240 / +110	0 / -52	+137 / +56	0 / -32	+69 / +17	0 / -32	+52 / 0	0 / -32	+16 / -36	0 / -32	-14 / -66	0 / -32	-36 / -88	0 / -32	-138 / -190
280	315	0 / -320	+650 / +330	0 / -130	+400 / +190	0 / -130	+240 / +110	0 / -52	+137 / +56	0 / -32	+69 / +17	0 / -32	+52 / 0	0 / -32	+16 / -36	0 / -32	-14 / -66	0 / -32	-36 / -88	0 / -32	-150 / -202
315	355	0 / -360	+720 / +360	0 / -140	+440 / +210	0 / -140	+265 / +125	0 / -57	+151 / +62	0 / -36	+75 / +18	0 / -36	+57 / 0	0 / -36	+17 / -40	0 / -36	-16 / -73	0 / -36	-41 / -98	0 / -36	-169 / -226
355	400	0 / -360	+760 / +400	0 / -140	+440 / +210	0 / -140	+265 / +125	0 / -57	+151 / +62	0 / -36	+75 / +18	0 / -36	+57 / 0	0 / -36	+17 / -40	0 / -36	-16 / -73	0 / -36	-41 / -98	0 / -36	-187 / -244
400	450	0 / -400	+840 / +440	0 / -155	+480 / +230	0 / -155	+290 / +135	0 / -63	+165 / +68	0 / -40	+83 / +20	0 / -40	+63 / 0	0 / -40	+18 / -45	0 / -40	-17 / -80	0 / -40	-45 / -108	0 / -40	-209 / -272
450	500	0 / -400	+880 / +480	0 / -155	+480 / +230	0 / -155	+290 / +135	0 / -63	+165 / +68	0 / -40	+83 / +20	0 / -40	+63 / 0	0 / -40	+18 / -45	0 / -40	-17 / -80	0 / -40	-45 / -108	0 / -40	-229 / -292

Clearance fits: C11/h11, D10/h9, E9/h9, F8/h7, G7/h6, H7/h6 — Transition fits: K7/h6, N7/h6 — Interference fits: P7/h6, S7/h6

Chapter 20
Geometrical tolerances and datum systems

Geometrical tolerances

The object of this section is to illustrate and interpret in simple terms the advantages of calling for geometrical tolerances on engineering drawings, and also to show that, when correctly used, they ensure that communications between the drawing office and the workshop are complete and incapable of mis-interpretation, regardless of any language barrier.

Applications

Geometrical tolerances are applied over and above normal dimensional tolerances when it is necessary to control more precisely the form or shape of some feature of a manufactured part, because of the particular duty that the part has to perform. In the past, the desired qualities would have been obtained by adding to drawings such expresions as 'surfaces to be true with one another', 'surfaces to be square with one another', 'surfaces to be flat and parallel', etc., and leaving it to workshop tradition to provide a satisfactory interpretation of the requirements.

Advantages

Geometrical tolerances are used to convey in a brief and precise manner complete geometrical requirements on engineering drawings. They should always be considered for surfaces which come into contact with other parts, especially when close tolerances are applied to the features concerned.

No language barrier exists, as the symbols used are in agreement with published recommendations of the International Organisation for Standardisation (ISO) and have been internationally agreed. BS 308: part 3 incorporates these symbols.

Caution It must be emphasized that geometrical tolerances should be applied only when real advantages result, when normal methods of dimensioning are considered inadequate to ensure that the design function is kept, especially where repeatability must be guaranteed. Indiscriminate use of geometrical tolerances could increase costs in manufacture and inspection. Tolerances should be as wide as possible, as the satisfactory design function permits.

General rules

The symbols relating to geometrical tolerances are shown in Fig. 20.1. Examination of the various terms – flatness, straightness, concentricity, etc. – used to describe the geometrical characteristics shows that one type of geometrical tolerance can control another form of geometrical error.

For example, a positional tolerance can control perpendicularity and straightness; parallelism, perpendicularity, and angularity tolerances can control flatness.

The use of geometrical tolerances does not involve or imply any particular method of manufacture or inspection. Geometrical tolerances shown in this book, in keeping with International conventions, must be met regardless of feature size unless modified by one of the following conditions:

a) Maximum material condition, denoted by the symbol Ⓜ describes a part which contains the maximum amount of material, i.e. the minimum size hole or the maximum size shaft.
b) Least material condition, denoted by the symbol Ⓛ describes a part which contains the minimum amount of material, i.e. the maximum size hole or the minimum size shaft.

Theorectically exact dimensions
(Fig. 20.2)

These dimensions are identified by enclosure in a rectangular box, e.g. 50 EQUI-SPACED 60° ⌀30 and are commonly known as 'Boxed dimensions'. They define the true position of a hole, slot, boss profile etc.

Boxed dimensions are never individually toleranced but are always accompanied by a positional or zone tolerance specified within the tolerance frame referring to the feature.

Definitions

Limits The maximum and minimum dimensions for a given feature are known as the 'limits'.

For example, 20 ± 0.1 or 20.1

19.9.

Type of tolerence	Characteristics to be toleranced	Symbol	Applications
Form	Straightness	—	A straight line. The edge or axis of a feature.
	Flatness	▱	A plane surface.
	Circularity (roundness)	○	The periphery of a circle. Cross-section of a bore, cylinder, cone or sphere.
	Cylindricity	⌭	The combination of circularity, straightness and parallelism of cylindrical surfaces. Mating bores and plungers.
	Profile of a line	⌒	The theoretical or perfect form of profiles defined by true boxed dimensions.
	Profile of a surface	⌓	The theoretical or perfect form of surfaces defined by true boxed dimensions.
Orientation	Parallelism	//	Parallelism of a feature related to a datum. Can control flatness when related to a datum.
	Perpendicularity (squareness)	⊥	Surfaces, axes, or lines positioned at right angles to each other.
	Angularity	∠	The angular displacement of surfaces, axes, or lines from a datum.
Location	Position	⊕	The deviation of a feature from a true position.
	Concentricity & coaxiality	◎	The relationship between two circles having a common centre or two cylinders having a common axis.
	Symmetry	≡	The symmetrical position of a feature related to a datum.
Runout	Circular runout	↗	The position of a point fixed on a surface of a part which is rotated 360° about its datum axis.
	Total runout	↗↗	The relative position of a point when traversed along a surface rotating about its datum axis.

Fig. 20.1

Fig. 20.2

The upper and lower limits of size are 20.1 mm and 19.9 mm respectively.

Tolerance The algebraic difference between the upper and lower limit of size is known as the 'tolerance'. In the example above, the tolerance is 0.2 mm. The tolerance is the amount of variation permitted.

Nominal dimension Limits and tolerances are based on 'nominal dimensions' which are target dimensions. In practice there is no such thing as a nominal dimension, since no part can be machined to an exact size, but the use of the term is an accepted convention.

The limits referred to above can be set in two ways:

a) *unilateral limits*—limits set wholly above or below the nominal size;
b) *bilateral limits*—limits set partly above and partly below the nominal size.

Geometrical tolerance These tolerances specify the maximum error of a component's geometrical shape, position, or size over its whole length or surface. This is done by defining a zone in which the feature may lie, and this may be modified only by a more restrictive explanatory note.

Tolerance zone A tolerance zone is an area or volume, enclosed by imaginary lines or surfaces, in which any deviation of the feature must be contained.

Method of indicating geometrical tolerances on drawings

Geometrical tolerances are indicated by stating the following details in compartments in a rectangular frame:

a) the characteristic symbol, for single or related features;
b) the tolerance value,
   i) preceded by $\varnothing$ if the zone is circular or cylindrical.
   ii) preceded by $S\varnothing$ if the zone is spherical;
c) letters identifying datum features when specified.

Fig. 20.3 shows examples.

Methods of applying the tolerance frame to the toleranced feature

Figs. 20.4 and 20.5 illustrate alternative methods of referring the tolerance to the surface or the plane itself. Note that in Fig. 20.5 the dimension line and frame leader line are offset.

Figs. 20.6 and 20.7 illustrate alternative methods of referring the tolerance to the axis or median plane. Note that in Fig. 20.7 the dimension line and frame leader line are drawn in line.

The tolerance frame as shown in Fig. 20.8 refers to the axis or median plane of the 'whole' part.

The tolerance frame as shown in Fig. 20.9 refers to the axis or median plane only of the dimensioned feature.

Left-hand compartment: symbol for characteristic

Third compartment: datum identification letter(s)

Second compartment: total tolerance value in the unit used for linear dimensions

Fig. 20.3

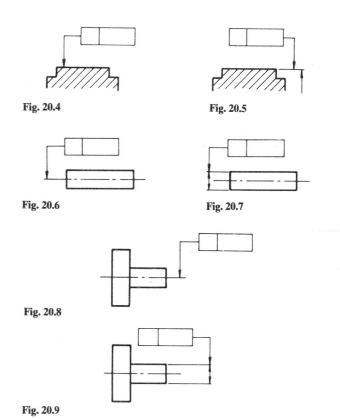

Fig. 20.4 **Fig. 20.5**

Fig. 20.6 **Fig. 20.7**

Fig. 20.8

Fig. 20.9

Procedure for positioning remarks which are related to tolerance

Remarks related to the tolerance for example '4 holes', '2 surfaces' or '6x' should be written above the frame.

4 HOLES
6×

Indications qualifying the feature within the tolerance zone should be written near the tolerance frame and may be connected by a leader line.

NOT CONVEX

If it is necessary to specify more than one tolerance characteristic for a feature, the tolerance specification should be given in tolerance frames positioned one under the other as shown.

The application of tolerances to a restricted length of a feature

Fig. 20.10 shows the method of applying a tolerance to only a particular part of a feature.

The tolerance frame in Fig. 20.11 shows the method of applying another tolerance, similar in type but smaller in magnitude, on a shorter length. In this case, the whole flat surface must lie between parallel planes 0.2 apart, but over any length of 180 mm, in any direction, the surface must be within 0.05.

Fig. 20.12 shows the method used to apply a tolerance over a given length; it allows the tolerance to accumulate over a longer length. In this case, a parallel tolerance of 0.02 is applicable over a length of 100 mm. If the total length of the feature was 800 mm, then the total permitted tolerance would accumulate to 0.16.

Fig. 20.10

Fig. 20.11

Fig. 20.12

Tolerance zones

The width of the tolerance zone is in the direction of the leader line arrow joining the symbol frame to the toleranced feature unless the tolerance zone is preceded by the symbol $\varnothing$.

Individual tolerance zones of the same value can be applied to more than one separate feature.

Projected toleranced zone

Fig. 20.14 shows a part section through a flange where it is required to limit the variation in perpendicularity of each hole axis. The method used is to apply a tolerance to a projected zone. The enlargement shows a possible position for the axis through one hole. Each hole axis must lie somewhere within a projected cylinder of $\varnothing$0.02 and 30 deep.

Fig. 20.13

Fig. 20.14

Note that the zone is projected from the specified datum.

Datums

A datum is in practice the functional plane or axis used for manufacture or inspection purposes, and should always be clearly identified on the drawing. A datum

Fig. 20.15

surface on a component should be accurately finished, since other locations or surfaces are established by measuring from the datum. Fig. 20.15 shows a datum surface indicated by the letter A.

In the above example, the datum edge is subject to a straightness tolerance of 0.05, shown in the tolerance frame. This datum is given a letter, to distinguish it from other datums, and the letter is boxed and connected to the tolerance frame by a leader line terminating in a solid equilateral triangle.

Methods of specifying datum features

Single datums Figs 20.16 and 20.17 illustrate alternative methods of positioning the datum box when the surface itself is the datum. Note that in fig. 20.17 the dimension line and datum box are offset.

Fig. 20.16   **Fig. 20.17**   Interpretation

Fig. 20.18 and 20.19 illustrate alternative methods of positioning the datum box when the axis or median plane is the datum. Note that in fig. 20.19 the dimension line and the datum box are in line.

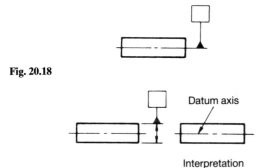

Fig. 20.18

Fig. 20.19

The datum box shown in Fig. 20.20 indicates that the whole length of the common axis or median plane of two or more features is the datum.

The datum box in Fig. 20.21 indicates that the datum axis or median plane refers only to the feature dimensioned.

Fig. 20.20   Interpretation

Fig. 20.21   Interpretation

Multiple datums Fig. 20.22 illustrates two datum features of equal status used to establish a single datum plane. The reference letters are placed in the third compartment of the tolerance frame, and have a hyphen separating them.

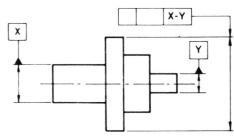

Fig. 20.22

Fig. 20.23 shows the drawing instruction necessary when the application of the datum features is required in a particular order of priority.

Fig. 20.23

Fig. 20.24 shows an application where a geometrical tolerance is related to two separate datum surfaces.

Functionally, both datums are of equal importance, and no priority is implied. Additional letters can be added in the third compartment if further datum references are required. No significance is attached to the letters used.

Fig. 20.24

Fig. 20.25

If a positional tolerance is required as shown in Fig. 20.25 and no particular datum is specified, then the individual feature to which the geometrical-tolerance frame is connected is chosen as the datum.

Fig. 20.26 illustrates a further positional-tolerance application where the geometrical reference frame is to be located in relation to another feature which may be indicated as the datum. In the given example, this implies that the pitch circle and the datum circle must be coaxial i.e. they have common axes.

Fig. 20.26

Datum targets

Surfaces produced by forging casting or sheet metal may be subject to bowing, warping or twisting; and not necessarily be flat. It is therefore impractical to designate an entire surface as a functional datum because accurate and repeatable measurements cannot be made from the entire surface.

In order to define a practical datum plane, selected points or areas are indicated on the drawing. Manufacturing processes and inspection utilizes these combined points or areas as datums.

Datum target symbols

The symbol for a datum target is a circle divided by a horizontal line. The lower part identifies the datum target. The upper area may be used only for information relating to datum target.

Indication of datum targets

If the datum target is:

a) a point: it is indicated by a cross .... **x**
b) a line: it is indicated by two crosses, connected by a thin line .... **x——x**
c) an area: it is indicated by a hatched area surrounded by a thin double dashed chain

All symbols appear on the drawing view which most clearly shows the relevant surface.

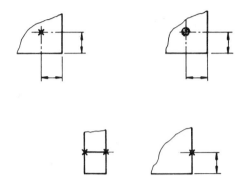

Practical application of datum targets

Fig. 20.27

Interpretation   It is understood from the above illustration that:

a) Datum targets A1, A2 and A3 establish Datum A.
b) Datum targets B1 and B2 establish Datum B.
c) Datum target C1 establishes Datum C.

Fig. 20.28

Fig. 20.28 illustrates the case where there is no functional reason to choose one or other of two associated features as a datum. The tolerance frame is applied as shown.

Fig. 20.29 illustrates the case where two associated features are chosen as a datum, from which another feature is toleranced.

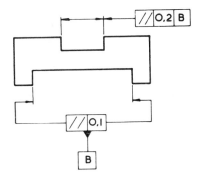

Fig. 20.29

Fig. 20.30 shows a datum for a positional tolerance which is derived from the true position of a group of holes.

Fig. 20.30

Fig. 20.31 shows a method of establishing the position of a datum from a group of geometrically toleranced features. Datum C is the axis of the group of six holes.

Datum C is derived by combining the following two conditions.

a) The 6 equally spaced holes in their theoretically exact positions are relative to primary datum A, with secondary datum B at its max. material condition.

b) The zone tolerance of the holes is ⌀0.01 when the holes are at the maximum material condition.

Fig. 20.31

Chapter 21
Application of geometrical tolerances

In this chapter, examples are given of the application of tolerances to each of the characteristics on engineering drawings by providing a typical specification for the product and the appropriate note which must be added to the drawing. In every example, the tolerance values given are only typical figures: the product designer would normally be responsible for the selection of tolerance values for individual cases.

Straightness

A straight line is the shortest distance between two points. A straightness tolerance controls:

1 the straightness of a line on a surface,
2 the straightness of a line in a single plane,
3 the straightness of an axis.

Case 1

Product requirement
The specified line shown on the surface must lie between two parallel straight lines 0.03 apart.

Drawing instruction
A typical application for this type of tolerance could be a graduation line on an engraved scale.

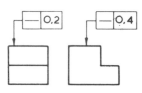

In the application shown above, tolerances are given controlling the straightness of two lines at right angles to one another. In the left-hand view the straightness control is 0.2, and in the right-hand view 0.4. As in the previous example, the position of the graduation marks would be required to be detailed on a plan view.

Case 2

Product requirement
The axis of the whole part must lie in a boxed zone of 0.3 × 0.2 over its length.

Drawing instruction
As indicated, the straightness of the axis is controlled by the dimensions of the box, and could be applied to a long rectangular key.

Case 3

Product requirement
The axis of the whole feature must lie within the cylindrical tolerance zone of 0.05.

Drawing instruction

Case 4

Product requirement
The geometrical tolerance may be required to control only part of the component. In this example the axis of the dimensioned portion of the feature must lie within the cylindrical tolerance zone of 0.1 diameter.

Drawing instruction

Flatness

Flatness tolerances control the divergence or departure of a surface from a true plane.

The tolerance of flatness is the specified zone between two parallel planes. It does not control the squareness or parallelism of the surface in relation to other features, and it can be called for independently of any size tolerance.

Case 1

Product requirement
The surface must be contained between two parallel planes 0.07 apart.

Drawing instruction

Case 2

Product requirement
The surface must be contained between two parallel planes 0.03 apart, but must not be convex.

Drawing instruction
Note that these instructions could be arranged to avoid a concave condition.

Circularity

Circularity is a condition where any point of a feature's continuous curved surface is equidistant from its centre, which lies in the same plane.

The tolerance of circularity controls the divergence of the feature, and the annular space between the two coplanar concentric circles defines the tolerance zone, the magnitude being the algebraic difference of the radii of the circles.

Case 1

Product requirement
The circumference of the bar must lie between two co-planar concentric circles 0.5 apart.

Drawing instruction
Note that, at any particular section, a circle may not be concentric with its axis but may still satisfy a circularity tolerance. The following diagram shows a typical condition.

Case 2

Product requirement
The circumference at any cross section must lie between two coplanar concentric circles 0.02 apart.

Drawing instruction

Case 3

Product requirement
The periphery of any section of maximum diameter of the sphere must lie between concentric circles a radial distance 0.001 apart in the plane of the section.

Drawing instruction

Cylindricity

The combination of parallelism, circularity and straightness defines cylindricity when applied to the surface of a cylinder, and is controlled by a tolerance of cylindricity. The tolerance zone is the annular space between two coaxial cylinders, the radial difference being the tolerance value to be specified.

It should be mentioned that, due to difficulties in checking the combined effects of parallelism, circularity and straightness, it is recommended that each of these characteristics is toleranced and inspected separately.

Product requirement
The whole curved surface of the feature must lie between an annular tolerance zone 0.04 wide formed by two cylindrical surfaces coaxial with each other.

Drawing instruction

Profile tolerance of a line

Profile tolerance of a line is used to control the ideal contour of a feature. The contour is defined by theoretically exact boxed dimensions and must be accompanied by a relative tolerance zone. This tolerance zone, unless otherwise stated, is taken to be equally disposed about the true form, and the tolerance value is equal to the diameter of circles whose centres lie on the true form. If it is required to call for the tolerance zone to be positioned on one side of the true form (i.e. unilaterally), the circumferences of the tolerance-zone circles must touch the theoretical contour of the profile.

Case 1

Product requirement
The profile is required to be contained within the bilateral tolerance zone.

Drawing instructions

Case 2

Product requirement
The profile is required to be contained within the unilateral tolerance zone.

Drawing instruction

The figure below shows an example where the toleranced profile of a feature has a sharp corner. The inner tolerance zone is considered to end at the intersection of the inner boundary lines, and the outer tolerance zone is considered to extend to the outer boundary-line intersections. Sharp corners such as these could allow considerable rounding; if this is desirable, then the design requirement must be clearly defined on the drawing by specifying a radius or adding a general note such as 'ALL CORNERS 0.5 MAX'. It should be noted that such radii apply regardless of the profile tolerance.

In the example given, the product is required to have a sharp corner.

Product requirement

Drawing instruction

Profile tolerance of a surface

Profile tolerance of a surface is used to control the ideal form of a surface, which is defined by theoretically exact boxed dimensions and must be accompanied by a relative tolerance zone. The profile-tolerance zone, unless otherwise stated, is taken to be bilateral and equally disposed about its true-form surface. The tolerance value is equal to the diameter of spheres whose centre lines lie on the true form of the surface. The zone is formed by surfaces which touch the circumferences of the spheres on either side of the ideal form.

If it is required to call for a unilateral tolerance zone, then the circumferences of the tolerance-zone spheres must touch the theoretical contour of the surface.

Product requirement
The tolerance zone is to be contained by upper and lower surfaces which touch the circumference of spheres 0.3 diameter whose centres lie on the theoretical form of the surface.

Drawing instruction

Parallelism

Two parallel lines or surfaces are always separated by a uniform distance. Lines or surfaces may be required to be parallel with datum planes or axes. Tolerance zones may be the space between two parallel lines or surfaces, or the space contained within a cylinder positioned parallel to its datum. The magnitude of the tolerance value is the distance between the parallel lines or surfaces, or the cylinder diameter.

Case 1

Product requirement
The axis of the hole on the left-hand side must be contained between two straight lines 0.2 apart, parallel to the datum axis X and lying in the same vertical plane.

Drawing instruction

Case 2

Product requirement
The axis of the upper hole must be contained between two straight lines 0.3 apart which are parallel to and symmetrically disposed about the datum axis X and lie in the same horizontal plane.

Drawing instruction

Case 3

Product requirement
The upper hole axis must be contained in a cylindrical zone 0.4 diameter, with its axis parallel to the datum axis X.

Drawing instruction

Case 4

Product requirement
The axis of the hole on the left-hand side must be contained in a tolerance box 0.5 × 0.2 × width, as shown, with its sides parallel to the datum axis X and in the same horizontal plane.

Drawing instruction

The tolerance zone is the space between two parallel lines or surfaces; it can also be the space contained within a cylinder. All tolerance zones are perpendicular to the datum feature.

The magnitude of the tolerance value is the specified distance between these parallel lines or surfaces, or the diameter of the cylinder.

Case 5

Product requirement
The axis of the hole must be contained between two planes 0.06 apart parallel to the datum surface X.

Drawing instruction

Case 6

Product requirement
The top surface of the component must be contained between two planes 0.7 apart and parallel to the datum surface X.

Drawing instruction

Perpendicularity (squareness)

Perpendicularity is the condition when a line, plane, or surface is at right angles to a datum feature.

Case 1

Product requirement
The axis of the vertical hole must be contained between two planes 0.1 apart which are perpendicular to the datum axis.

Drawing instruction

Case 2

Product requirement
The axis of the upright must be contained between two straight lines 0.2 apart which are perpendicular to the datum. Squareness is controlled here in one plane only.

Drawing instruction

Case 3

Product requirement
The axis of the column must be contained in a cylindrical tolerance zone 0.3 diameter, the axis of which is perpendicular to the datum surface X. Squareness is controlled in more than one plane by this method.

Drawing instruction

Drawing instruction

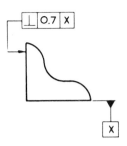

Case 4

Product requirement
The axis of the column must be contained in a tolerance-zone box 0.2 × 0.4 which is perpendicular to the datum surface X.

Drawing instruction

Case 5

Product requirement
The left-hand end face of the part must be contained between two parallel planes 0.8 apart and perpendicular to the datum axis X.

Drawing instruction

Case 6

Product requirement
The left-hand surface must be contained between two parallel planes 0.7 apart and perpendicular to the datum surface X.

Angularity

Angularity defines a condition between two related planes, surfaces, or lines which are not perpendicular or parallel to one another. Angularity tolerances control this relationship.

The specified angle is a basic dimension, and is defined by a theoretically exact boxed dimension and must be accompanied by a tolerance zone. This zone is the area between two parallel lines inclined at the specified angle to the datum line, plane, or axis. The tolerance zone may also be the space within a cylinder, the tolerance value being equal to the cylinder diameter: In this case, symbol ∅ precedes tolerance value in the tolerance frame.

Case 1

Product requirement
The inclined surface must be contained within two parallel planes 0.2 apart which are at an angle of 42° to the datum surface.

Drawing instruction

Case 2

Product requirement
The axis of the hole must be contained within two parallel straight lines 0.1 apart inclined at 28° to the datum axis.

Drawing instruction

Case 3

Product requirement
The inclined surface must be contained within two parallel planes 0.5 apart which are inclined at 100° to the datum axis.

Drawing instruction

Case 1

Product requirement
The circular radial run-out must not exceed 0.4 at any point along the cylinder, measured perpendicular to the datum axis without axial movement.

Drawing instruction

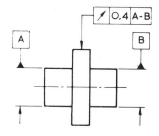

Case 2

Product requirement
Circular run-out must not exceed 0.2 measured at any point normal to the surface, without axial movement.

Drawing instruction

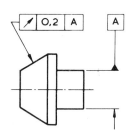

Circular run-out

Circular run-out is a unique geometrical tolerance. It can be a composite form control relating two or more characteristics, and it requires a practical test where the part is rotated through 360° about its own axis.

The results of this test may include errors of other characteristics such as circularity, concentricity, perpendicularity, or flatness, but the test cannot discriminate between them. It should therefore not be called for where the design function of the part necessitates that the other characteristics are to be controlled separately. The sum of any of these errors will be contained within the specified circular run-out tolerance value. The tolerance value is equal to the full indicator movement of a fixed point measured during one revolution of the component about its axis, without axial movement. Circular run-out is measured in the direction specified by the arrow at the end of the leader line which points to the toleranced feature. It must always be measured regardless of feature size, and it is convenient for practical purposes to establish the datum as the diameter(s) from which measurement is to be made, if true centres have not been utilised in manufacturing the part.

Case 3

Product requirement
At any radius, the circular run-out must not exceed 0.06 measured parallel to the datum axis.

Drawing instruction

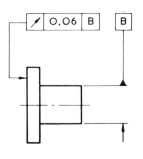

Case 4

Product requirement

The component is required to be rotated about datum axis C, with datum face B set to ensure no axial movement.

The circular radial runout on the cylindrical portion must not exceed 0.05 at any point measured perpendicular to the datum axis.

The circular runout on the tapered portion must not exceed 0.07 at any point measured normal to its surface.

The circular runout on the curved portion must not exceed 0.04 at any point measured normal to its surface.

The axial runout of the end face must not exceed 0.1 at any point measured parallel to the datum axis of rotation.

DATUM FACE B

DATUM AXIS C

ROTATE PART ABOUT DATUM AXIS

Drawing instruction

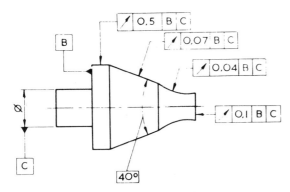

B

C

0.5 B C

0.07 B C

0.04 B C

0.1 B C

40°

Circular runout provides composite control of circular elements of a surface.

Total runout provides composite control of all surface elements. The complete surface is measured, and not single points, as in circular runout.

Total runout controls cumulative variations of perpendicularity which can detect wobble, also flatness which can detect concavity and convexity.

Total runout

Case 1

Product requirement

The total runout must not exceed 0.06 at any point measured across the entire surface parallel to the datum axis.

ROTATE PART ABOUT DATUM AXIS

Drawing instruction

0.06 B B

Note. The symbol $\diagup\!\!/$ means that the measuring instrument is guided across a theoretically exact surface true to the datum axis.

Position

A positional tolerance controls the location of one feature from another feature or datum.

The tolerance zone can be the space between two parallel lines or planes, a circle, or a cylinder. The zone defines the permissible deviation of a specified feature from a theoretically exact position.

The tolerance value is the distance between the parallel lines or planes, or the diameters of the circle or cylinder.

The theoretically exact position also incorporates squareness and parallelism of the tolerance zones with the plane of the drawing.

Case 1

Product requirement

The point must be contained within a circle of 0.1 diameter in the plane of the surface. The circle has its centre at the intersection of the two theoretically exact dimensions. If the point were to be located by three dimensions, the tolerance zone would be a sphere.

TOLERANCE ZONE 0.1 DIA.

Drawing instruction

80

70

⊕ Ø0.1

Case 2

Product requirement

The axis of the hole must be contained in a cylindrical tolerance zone of 0.3 diameter with its axis coincident with the theoretically exact position of the hole axis.

TOLERANCE ZONE 0.3 DIA.

Drawing instruction

Drawing instruction

Case 3

Product requirement
Each line must be contained between two parallel straight lines on the surface, 0.2 apart, which are symmetrical with the theoretically exact positions of the required lines.

Drawing instruction

Case 4

Product requirement
The axes of each of the four holes must be contained in a cylindrical tolerance zone of 0.5 diameter, with its own axis coincident with the theoretically exact position of each hole.

Drawing instruction

Case 5

Product requirement
The axes of each of the four holes must be contained in a boxed zone of 0.04 × 0.03 × 10, symmetrically disposed about the theoretically exact position of each hole axis.

Case 6

Product requirement
The angled surface must be contained between two parallel planes 0.7 apart, which are symmetrically disposed about the theoretically exact position of the surface relative to datum axis X and datum face Y.

Drawing instruction

Case 7

Product requirement
The axes of the two holes must be contained in cylindrical tolerance zones of 0.01 diameter, with their own axes coincident with the theoretically exact hole positions related to datum face X and the datum centre-line axis Y.

Drawing instruction

Concentricity and coaxiality

Two circles are said to be concentric when their centres are coincident.

Two cylinders are said to be coaxial when their axes are coincident.

The deviation from the true centre or datum axis is controlled by the magnitude of the tolerance zone.

Case 1 (Concentricity)

Product requirement
To contain the centre of the large circle within a circular tolerance zone of 0.001 diameter which has its centre coincident with the datum-circle centre.

Drawing instruction

Case 2 (Coaxiality)

Product requirement
To contain the axis of the right-hand cylinder within a cylindrical tolerance zone which is coaxial with the axis of the datum cylinder.

Drawing instruction

Case 3 (Coaxiality)

Product requirement
To contain the axes of both the left- and right-hand cylinders within a cylindrical tolerance zone.

Drawing instruction

Case 4 (Coaxiality)

Product requirement
To contain the axis of the central cylinder within a cylindrical tolerance zone which is coaxial with the mean axes of the left- and right-hand cylinders.

Drawing instruction

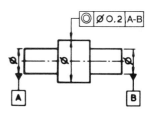

Symmetry

Symmetry involves the division of spacing of features so that they are positioned equally in relation to a datum which may be a line or plane. The tolerance zone is the space between two parallel lines or planes, parallel to, and positioned symmetrically with the datum. The tolerance magnitude is the distance between these two parallel lines or planes.

Symmetry also implies perpendicularity with the plane of the drawing where depth is involved.

Case 1

Product requirement
The specified line XX must lie in a tolerance zone formed by two parallel straight lines 0.01 apart and disposed symmetrically between datums A and B.

Drawing instruction

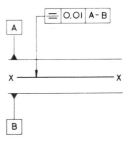

Case 2

Product requirement
The axis of a hole in a plate must lie in a rectangular-box tolerance zone $0.03 \times 0.06 \times$ depth of the plate, parallel with and symmetrically disposed about the common median planes formed by slots AC and BD.

Drawing instruction

Case 3

Product requirement
The median plane of the slot and tongue must lie between two parallel planes 0.4 apart.

Drawing instruction

Chapter 22
Maximum material and least material principles

Maximum material condition (MMC)

MMC is that condition of a part or feature which contains the maximum amount of material, e.g. minimum size hole, or a maximum size shaft. In certain cases it's use allows an increase in the specified tolerance if it is indicated that the tolerance applies to the feature at its maximum material condition.

The *maximum material principle* takes into account the mutual dependence of tolerances of size, form, orientation and/or location and permits additional tolerance as the considered feature departs from it's maximum material condition.

The free assembly of components depends on the combined effect of the actual finished sizes and the errors of form or position of the parts.

Any errors of form or position between two mating parts have the effect of virtually altering their respective sizes. The tightest condition of assembly between two mating parts occurs when each feature is at the MMC, plus the maximum errors permitted by any required geometrical tolerance.

Assembly clearance is increased if the actual sizes of the mating features are finished away from their MMC, and if any errors of form or position are less than that called for by any geometrical control. Also, if either part is finished away from its MMC, the clearance gained could allow for an increased error of form or position to be accepted. Any increase of tolerance gained this way, provided it is functionally acceptable for the design, is advantageous during the process of manufacture.

The symbol for maximum material condition is the letter M enclosed by a circle, Ⓜ. The symbol is positioned in the tolerance frame as follows:

(a) [⊕ | ⌀ 0.05Ⓜ | X]   Refers to the tolerance only.

(b) [⊕ | ⌀0.05 | X Ⓜ]   Refers to the datum only.

(c) [⊕ | ⌀ 0.05Ⓜ | X Ⓜ]   Refers to both tolerance and datum.

Least material condition (LMC)

LMC is that condition of a part or feature which contains the minimum amount of material, e.g. maximum size hole or a minimum size shaft.

Circumstances do arise where, for example, a designer would require to limit the minimum wall thickness between a hole and the side of a component. In such a case we need to control the least material condition where a part contains the minimum amount of material.

The appropriate tolerance would then be quoted, followed by the letter L inside a circle Ⓛ.

Generally such examples are very few. The applications which follow cover the more widely found conditions of MMC.

Maximum material condition related to geometrical form

The limit of size, together with geometrical form or position of a feature, are factors of the maximum material principle, and its application is restricted to those features whose size is specified by toleranced dimensions incorporating an axis or median plane. It can never be applied to a plane, surface, or line on a surface.

STRAIGHTNESS ERROR

ACTUAL DIAMETER

ACTUAL DIAMETER

VIRTUAL SIZE BEING THE EFFECTIVE ASSEMBLY DIAMETER

Fig. 22.1

The characteristics to which it can be applied are as follows:

> straightness,
> parallelism,
> squareness,
> angularity,
> position,
> concentricity,
> symmetry

The characteristics to which the maximum material condition concept cannot be applied are as follows:

> flatness,
> roundness,
> cylindricity.
> profile of a line,
> profile of a surface,
> run-out.

Maximum material condition applied to straightness

Fig. 22.2 shows a typical drawing instruction where limits of size are applied to a pin, and in addition a straightness tolerance of 0.2 is applicable at the maximum material condition.

Fig. 22.2

Fig. 22.3 shows the condition where the pin is finished at the maximum material condition with the maximum straightness error. The effective assembly diameter will be equal to the sum of the upper limit of size and the straightness tolerance.

The straightness error is contained within a cylindrical tolerance zone of ∅0.2.

Fig. 22.3

To provide the same assembly diameter of 20.7 as shown in fig. 22.4 when the pin is finished at its low limit of size of 20.0, it follows that a straightness error of 0.7 could be acceptable. This increase may in some cases have no serious effect on the function of the component, and can be permitted.

Fig. 22.4

Maximum material condition applied to squareness

Fig. 22.5 shows a typical drawing instruction where limits of size are applied to a pin, and in addition a squareness tolerance of 0.3 is applicable at the maximum material condition.

Fig. 22.5

Fig. 22.6 shows the condition where the pin is finished at the maximum material condition with the maximum squareness error of 0.3. The effective assembly diameter will be the sum of the upper limit of size and the squareness error. The squareness error will be contained within a cylindrical tolerance zone of ∅0.3.

Fig. 22.6

To provide the same assembly diameter of 30.9, as shown in Fig. 22.7, when the pin is finished at its low limit of size of 30.0, it follows that the squareness error could increase from 0.3 to 0.9. This permitted increase should be checked for acceptability.

Fig. 22.7

Maximum material condition applied to position

A typical drawing instruction is given in Fig. 22.8, and the following illustrations show the various extreme dimensions which can possibly arise.

Fig. 22.8

Condition A (Fig. 22.9)
Minimum distance between hole centres and the maximum material condition of holes.

Fig. 22.9

Condition B (Fig. 22.10)
Maximum distance between hole centres and maximum material condition of holes

Fig. 22.10

Condition C (Fig. 22.11)
To give the same assembly condition as in A, the minimum distance between hole centres is reduced when the holes are finished away from the maximum material condition.

Fig. 22.11

Condition D (Fig. 22.12)
To give the same assembly condition as in B, the maximum distance between hole centres is increased when the holes are finished away from the maximum material condition.

Note that the total tolerance zone is 0.2 + 0.1 = 0.3, and therefore the positional tolerance can be increased where the two holes have a finished size away from the maximum material condition.

Fig. 22.12

Maximum material condition applied to coaxiality

In the previous examples, the geometrical tolerance has been related to a feature at its maximum material condition, and, provided the design function permits, the tolerance has increased when the feature has been finished away from the maximum material condition. Now the geometrical tolerance can also be specified in relation to a datum feature, and Fig. 22.13 shows a typical application and drawing instruction of a shoulder on a shaft. The shoulder is required to be coaxial with the shaft, which acts as the datum. Again, provided the design function permits, further relaxation of the quoted geometrical control can be achieved by applying the maximum material condition to the datum itself.

Fig. 22.13

Various extreme combinations of size for the shoulder and shaft can arise, and these are given in the drawings below. Note that the increase in coaxiality error which could be permitted in these circumstances is equal to the total amount that the part is finished away from its maximum material condition, i.e. the shoulder tolerance plus the shaft tolerance.

Condition A (Fig. 22.14). Shoulder and shaft at maximum material condition; shoulder at maximum permissible eccentricity to the shaft datum axis X.

Fig. 22.14

Condition B (Fig. 22.15). Shoulder at minimum material condition and shaft at maximum material condition. Total coaxiality tolerance = specified coaxiality tolerance + limit of size tolerance of shoulder = 0.2 + 0.2 = 0.4 diameter. This gives a maximum eccentricity of 0.2.

Fig. 22.15

Condition C (Fig. 22.16). shows the situation where the smallest size shoulder is associated with the datum shaft at its low limit of size. Here the total coaxiality tolerance which may be permitted is the sum of the specified coaxiality tolerance + limit of size tolerance for the shoulder + tolerance on the shaft = 0.2 + 0.2 + 0.02 = 0.42 diameter.

Fig. 22.16

Maximum material condition and perfect form

When any errors of geometrical form are required to be contained within the maximum material limits of size, it is assumed that the part will be perfect in form at the upper limit of size.

In applying this principle, two conditions are possible.

Case 1 The value of the geometrical tolerance can progressively increase provided that the assembly diameter does not increase above the maximum material limit of size. Fig. 22.17 shows a shaft and the boxed dimension, and indicates that at maximum material limit of size the shaft is required to be perfectly straight.

Fig. 22.17

Fig. 22.18 shows the shaft manufactured to its lower limit of size, where the permitted error in straightness can be 0.05, since the assembly diameter will be maintained at 16.00. Similarly, a shaft manufactured to, say, 15.97 can have a permitted straightness error of 0.03.

Fig. 22.18

Case 2 The geometrical tolerance can also be limited to a certain amount where it would be undesirable for the part to be used in service too much out of line.

Fig. 22.19 shows a shaft, and the boxed dimension indicates that at the maximum material limit of size the shaft is required to be perfectly straight. Also, the upper part of the box indicates that a maximum geometrical tolerance error of 0.02 can exist, provided that for assembly purposes the assembly diameter does not exceed 14.00.

Fig. 22.19

Fig. 22.20 shows the largest diameter shaft acceptable, assuming that it has the full geometrical error of 0.02. Note that a shaft finished at 13.99 would be permitted a maximum straightness error of only 0.01 to conform with the drawing specification.

Fig. 22.20

Fig. 22.21 shows the smallest diameter shaft acceptable, and the effect of the full geometrical error of straightness.

Fig. 22.21

The application of maximum material condition and its relationship with perfect form and squareness

A typical drawing instruction is shown in Fig. 22.22.

Fig. 22.22

Condition A (Fig. 22.23). Maximum size of feature: zero geometrical tolerance.

Fig. 22.23

Condition B (Fig. 22.24). Minimum size of feature; Permitted geometrical error = 0.6.

Note that between these extremes the geometrical tolerance will progressively increase; i.e. when the shaft diameter is 30.3, then the cylindrical tolerance error permitted will be 0.3.

Fig. 22.24

The application of maximum material condition and its relationship with perfect form and coaxiality

A typical drawing instruction is shown in Fig. 22.25.

Fig. 22.25

Condition A (Fig. 22.26). Head and shank at maximum material condition. No geometrical error is permitted, and the two parts of the component are coaxial.

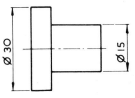

Fig. 22.26

Condition B (Fig. 22.27). Head at maximum material condition; shank at minimum material condition. The permitted geometrical error is equal to the tolerance on the shank size. This gives a tolerance zone of 0.1 diameter.

Fig. 22.27

Condition C (Fig. 22.28). Shank at maximum material condition; head at minimum material condition. The permitted geometrical error is equal to the tolerance on the head size. This gives a tolerance zone of 0.1 diameter.

Fig. 22.28

Condition D (Fig. 22.29). Both shank and shaft are finished at their low limits of size; hence the permitted geometrical error will be the sum of the two manufacturing tolerances, namely 0.2 diameter.

Fig. 22.29

The application of maximum material condition to two mating components

Fig. 22.30 shows a male and female component dimensioned with a linear tolerance between centres, and which will assemble together under the most adverse conditions allowed by the specified tolerances. The male component has centre distance and diameters of pins at maximum condition. The female component has centre distance and diameter of holes at minimum condition.

Fig. 22.30

The tolerance diagram, Fig. 22.31, shows that, when the pin diameters are at the least material condition, their centre distance may vary between $74.9 - 14.7 = 60.2$, or $45.1 + 14.7 = 59.8$. Now this increase in tolerance can be used to advantage, and can be obtained by applying the maximum material concept to the drawing detail.

Fig. 22.31

Similarly, by applying the same principle to the female component, a corresponding advantage is obtainable. The lower part of Fig. 22.31 shows the female component in its maximum material condition. Assembly with the male component will be possible if the dimension over the pins does not exceed 74.9 and the dimension between the pins is no less than 45.1.

Fig. 22.32 shows the method of dimensioning the female component with holes controlled by a positional tolerance, and modified by maximum material condition. This ensures assembly with the male component, whose pins are manufactured regardless of feature size.

When the maximum condition is applied to these features, any errors of form or position can be checked by using suitable gauges.

Fig. 22.32

Chapter 23
Coordinate and true positional tolerances

The holes in the component shown in Fig. 23.1 are positioned along one axis by toleranced co-ordinate dimensions. Since the hole centres between each pair of holes are toleranced separately, it follows that the limits are cumulative, e.g. between holes A and B they are ± 0.1, between holes A and C ± 0.2, and between holes A and D ± 0.3.

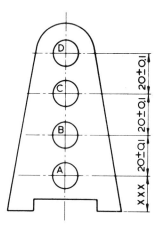

Fig. 23.1

Fig. 23.2 shows an alternative method of applying the co-ordinate method, from a common datum which is the centre line of hole A. Note that the distances between holes B and C could be either 20.2 or 19.8, i.e. 20 ± 0.2, depending on whether holes B and C are

drilled at their high or low limits of distance with respect to hole A. Similarly, between holes B and D the distance could be 40.2 or 39.8, i.e. 40 ± 0.2.

No further accumulation of error results after the first pair of hole pitches if this method is used on a long chain of holes; nevertheless, if the acceptable practical error is only ± 0.1, then the drawing must have limits of ± 0.05 applied for the above reason.

A further example of toleranced co-ordinate dimensioning is shown in Fig. 23.3, where two holes are positioned from datum edges.

Fig. 23.3

This type of dimensioning assumes that the two datum edges are exactly at 90° to each other. Should this not be the case, then the distance between the co-ordinate tolerances zones shown in Fig. 23.4 could vary considerably when the angle is greater or less than 90°.

Fig. 23.2

Fig. 23.4

True-position (theoretical exact) dimensioning

True-position dimensioning defines the exact location on a component of features such as holes, slots, keyways, etc., and also differentiates between 'ideal' and other toleranced dimensions. True-position dimensions are always shown 'boxed' on engineering drawings; they are never individually toleranced, and must always be accompanied by a positional or zone tolerance for the feature to which they are applied.

The positional tolerance is the permitted deviation of a feature from a true position.

The positional-tolerance zone defines the region which contains the extreme limits of position and can be, for example, rectangular, circular, cylindrical, etc.

Typical product requirement

In the example shown in Figs 23.5 and 23.6, the hole axis must lie within the cylindrical tolerance zone fixed by the true-position dimensions.

Fig. 23.5   Product requirement

Fig. 23.6   Drawing instruction

Some advantages of using this method are:

1   interpretation is easier, since true boxed dimensions fix the exact positions of details;
2   there are no cumulative tolerances;
3   it permits the use of functional gauges to match the mating part;
4   it can ensure interchangeability without resorting to small position tolerances, required by the co-ordinate tolerancing system;

5   the tolerancing of complicated components is simplified;
6   positional-tolerance zones can control squareness and parallelism.

The following examples show some typical cases where positional tolerances are applied to engineering drawings.

Case 1 (Figs 23.7 and 23.8).   The axes of the four fixing holes must be contained within cylindrical tolerance zones 0.03 diameter.

Fig. 23.7   Case 1: Product requirement

Fig. 23.8   Case 1: Drawing instruction

Case 2 (Figs 23.9 and 23.10).   The axes of the four fixing holes must be contained within rectangular tolerance zones 0.04×0.02.

Fig. 23.9   Case 2: Product requirement

Fig. 23.10 Case 2: Drawing instruction

Case 3 Fig. 23.11 shows a component where the outside diameter at the upper end is required to be square and coaxial within a combined tolerance zone with face A and diameter B as the primary and secondary datums.

Fig. 23.11 Case 3

Case 4 In the component illustrated in Fig. 23.12, the three dimensioned features are required to be perfectly square to the datum face A, and also truly coaxial with each other in the maximum material condition.

Fig. 23.12 Case 4

Case 5 (Figs 23.13 and 23.14). The six bolt-holes on the flange in Fig. 23.13 must have their centres positioned within six tolerance zones of ⌀0.25 when the bolt holes are at their maximum material condition (i.e. minimum limit of size).

Fig. 23.13 Case 5: Product requirement

Note in Fig. 23.14 that all the features in the group have the same positional tolerance in relation to each other. This method also limits in all directions the relative displacement of each of the features to each other.

Fig. 23.14 Case 5: Drawing instruction

Case 6 (Figs 23.15 and 23.16). The group of holes in Fig. 23.15, dimensioned with a positional tolerance, is also required to be positioned with respect to the datum spigot and the face of the flange.

Note in Fig. 23.16 that the four holes and the spigot are dimensioned at the maximum material condition. It

4 cylindrical tolerance zones 0.25 dia in true position relative to Datum Face B and Datum Axis A

Ø80

Fig. 23.15   Case 6: Product requirement

12.25
4 HOLES Ø12.00 EQUISPACED

Fig. 23.16   Case 6: Drawing instruction

follows that, if any hole is larger than 12.00, it will have the effect of increasing the positional tolerance for that hole. If the spigot is machined to less than 50.05, then the positional tolerance for the four holes as a group will also increase.

Case 7   Fig. 23.17 shows a drawing instruction where the group of equally spaced holes is required to be positioned relative to a coaxial datum bore.

7.1
6 HOLES Ø7.0 EQUISPACED

50.5
Ø49.5

Fig. 23.17   Case 7

Case 8   In the example in Fig. 23.18, control of the two Ø10 holes is required with respect to the datum hole X. The drawing instruction implies that the two Ø10 holes must be in line, dictated by the boxed dimension, but subject to a positional tolerance of Ø0.3. In addition, their centres are subject to variation by virtue of the positional tolerance given to the Ø16 datum hole. Fig 34.19 shows an extreme but acceptable condition for the position of the hole centres.

10.01
2 HOLES Ø10.00

16,1
1 HOLE Ø16,0

Fig. 23.18   Case 8

Cylindrical tolerance zone of Datum hole

The centres of the cylindrical tolerance zones of the two holes must be contained within the cylindrical tolerance zone established by the datum

Fig. 23.19   Case 8: Extreme condition

Case 9   The example in Fig. 23.20 shows three holes, of two different sizes, subject to the same positional tolerance. Since co-ordinate toleranced dimensions fix the position of the first hole, and true boxed dimensions fix the distance between holes, the extreme but acceptable condition shown in Fig. 23.21 can arise.

Case 10   The situation in case 9 can be extended to a larger group of holes fixed by boxed dimensions, as Fig. 23.22 illustrates. The extreme but acceptable condition shown in Fig. 23.23 can arise.

Fig. 23.20 Case 9

Fig. 23.22 Case 10

Locating square
co-ordinate
tolerance zone

The centres of the cylindrical positional
tolerance zone must be contained within
the square tolerance zone

Fig. 23.21 Case 9: Extreme condition

Locating square
co-ordinate
tolerance zone

The centres of the cylindrical
positional tolerance zones
must be contained within the
square tolerance zones

Fig. 23.23 Case 10: Extreme condition

Chapter 24
Cams and gears

A cam is generally a disc or a cylinder mounted on a rotating shaft, and it gives a special motion to a *follower*, by direct contact. The cam profile is determined by the required follower motion and the design of the type of follower.

The motions of cams can be considered to some extent as alternatives to motions obtained from linkages, but they are generally easier to design, and the resulting actions can be accurately predicted. If, for example, a follower is required to remain stationary, then this is achieved by a concentric circular arc on the cam. For a specified velocity or acceleration, the displacement of the follower can easily be calculated, but these motions are very difficult to arrange precisely with linkages.

Specialist cam-manufacturers computerise design data and, for a given requirement, would provide a read-out with cam dimensions for each degree, minute, and second of camshaft rotation.

When used in high-speed machinery, cams may require to be balanced, and this becomes easier to perform if the cam is basically as small as possible. A well-designed cam system will involve not only consideration of velocity and acceleration but also the effects of out-of-balance forces, and vibrations. Suitable materials must be selected to withstand wear and the effect of surface stresses.

Probably the most widely used cam is the *plate cam*, with its contour around the circumference. The line of action of the follower is usually either vertical or parallel to the camshaft, and Fig. 24.1 shows several examples.

Examples are given later of a *cylindrical* or *drum cam*, where the cam groove is machined around the circumference, and also a *face cam*, where the cam groove is machined on a flat surface.

Cam followers

Various types of cam followers are shown in Fig. 24.1.

Knife-edge followers are restricted to use with slow-moving mechanisms, due to their rapid rates of wear. Improved stability can be obtained from the roller follower, and increased surface area in contact with the cam can be obtained from the flat and mushroom types of follower. The roller follower is the most expensive type, but is ideally suited to high speeds and applications where heat and wear are factors.

Fig. 24.1 Plate cams (a) Plate cam with knife-edge follower (b) Plate cam with roller follower (c) Plate cam with flat follower (d) Plate cam with oscillating knife-edge follower (e) Plate cam with oscillating flat follower

Cam follower motions

1 Uniform velocity This motion is used where the follower is required to rise or fall at a constant speed, and is often referred to as 'straight-line' motion. Part of a uniform-velocity cam graph is shown in Fig. 24.2.

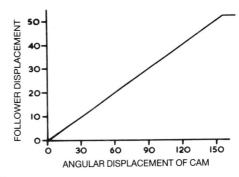

Fig. 24.2

Abrupt changes in velocity with high-speed cams result in large accelerations and cause the followers to jerk or chatter. To reduce the shock on the follower, the cam graph can be modified as indicated in Fig. 24.3 by adding radii to remove the sharp corners. However, this action results in an increase in the average rate of rise or fall of the follower.

Fig. 24.3

2 Uniform acceleration and retardation

motion is shown in Fig. 24.4. The graphs for both parts of the motion are parabolic. The construction for the parabola involves dividing the cam-displacement angle into a convenient number of parts, and the follower displacement into the same number of parts. Radial lines are drawn from the start position to each of the follower division lines, and the parabola is obtained by drawing a line through successive intersections. The uniform-retardation parabola is constructed in a similar manner, but in the reverse position.

Fig. 24.4

3 Simple harmonic motion is shown in

Fig. 24.5 where the graph is a sine curve. The construction involves drawing a semi-circle and dividing it into the same number of parts as the cam-displacement angle. The diameter of the semi-circle is equal to the rise or fall of the follower. The graph passes through successive intersections as indicated.

The application of the various motions to different combinations of cams and followers is shown by the following practical examples.

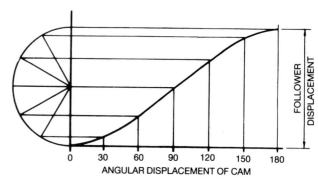

Fig. 24.5

Case 1 (Fig. 24.6)

Cam specification:
Plate cam, rotating anticlockwise. Point follower. Least radius of cam, 30 mm. Camshaft diameter, 20 mm.
0–90°, follower rises 20 mm with uniform velocity.
90–150°, follower rises 30 mm with simple harmonic motion.
150–210°, dwell period.
210–270°, follower falls 20 mm with uniform acceleration.
270–360°, follower falls 30 mm with uniform retardation.

Fig. 24.6

Fig. 24.7

1 Draw the graph as shown. Exact dimensions are used for the *Y* axis, where the follower lift is plotted. The *X* axis has been drawn to scale, where 12 mm represents 30° of shaft rotation.
2 To plot the cam, draw a 20 mm diameter circle to represent the bore for the camshaft, and another circle 30 mm radius to represent the base circle, or the least radius of the cam, i.e. the nearest the follower approaches to the centre of rotation.
3 Draw radial lines 30° apart from the cam centre, and number them in the reverse direction to the cam rotation.
4 Plot the *Y* ordinates from the cam graph along each of the radial lines in turn, measuring from the base circle. Where rapid changes in direction occur, or where there is uncertainty regarding the position of the profile, more points can be plotted at 10° or 15° intervals.
5 Draw the best curve through the points to give the required cam profile.

Note. The user will require to know where the cam program commences, and the zero can be conveniently established on the same centre line as the shaft keyway. Alternatively, a timing hole can be drilled on the plate, or a mark may be engraved on the plate surface. In cases where the cam can be fitted back to front, the direction of rotation should also be clearly marked.

Case 2 (Fig. 24.7)

Cam specification:
Plate cam, rotating anticlockwise. Flat follower. Least distance from follower to cam centre, 30 mm. Camshaft diameter, 20 mm.
0–120°, follower rises 30 mm with uniform velocity (modified).
120–210°, dwell period.
210–360°, follower falls 30 mm with uniform velocity (modified).

1 Draw the cam graph as shown, and modify the curve to remove the sharp corners. Note that in practice the size of the radius frequently used here varies between one-third and the full lift of the follower for the uniform-velocity part of the graph; the actual value depends on the rate of velocity and the speed of rotation. This type of motion is not desirable for high speeds.
2 Draw the base circle as before. 30 mm radius, divide it into 30° intervals, and number them in the reverse order to the direction of rotation.
3 Plot the *Y* ordinates from the graph, radially from the base circle along each 30° interval line.
 Draw a tangent at each of the plotted points, as shown, and draw the best curve to touch the tangents. The tangents represent the face of the flat follower in each position.
4 Check the point of contact between the curve and each tangent and its distance from the radial line. Mark the position of the widest point of contact. In

the illustration given, point P appears to be the greatest distance, and hence the follower will require to be at least *R* in radius to keep in contact with the cam profile at this point. Note also that a flat follower can be used only where the cam profile is always convex.

Although the axis of the follower and the face are at 90° in this example, other angles are in common use.

Case 3 (Fig. 24.8)

Cam specification:
Plate cam, rotating clockwise. 20 mm diameter roller follower.
30 mm diameter camshaft. Least radius of cam, 35 mm.
0–180°, rise 64 mm with simple harmonic motion.
180–240°, dwell period.
240–360°, fall 64 mm with uniform velocity.

1 Draw the cam graph as shown.
2 Draw a circle (shown as RAD Q) equal to the least radius of the cam plus the radius of the roller, and divide it into 30° divisions. Mark the camshaft angles in the anticlockwise direction.
3 Along each radial line plot the *Y* ordinates from the graph, and at each point draw a 20 mm circle to represent the roller.
4 Draw the best profile for the cam so that the cam touches the rollers tangentially, as shown.

Fig. 24.8

Fig. 24.9

Case 4 (Fig. 24.9)

Cam specification:
Plate cam, rotating clockwise. 20 mm diameter roller follower set 20 mm to the right of the centre line for the camshaft. Least distance from the roller centre to the camshaft centre line, 50 mm. 25 mm diameter camshaft.
0–120°, follower rises 28 mm with uniform acceleration.
120–210°, follower rises 21 mm with uniform retardation.
210–240°, dwell period.
240–330°, follower falls 49 mm with uniform velocity.
330–360°, dwell period.

1 Draw the cam graph as shown.
2 Draw a 20 mm radius circle, and divide it into 30° divisions as shown.
3 Where the 30° lines touch the circumference of the 20 mm circle, draw tangents at these points.
4 Draw a circle of radius Q, as shown, from the centre of the camshaft to the centre of the roller follower.

This circle is the base circle.
5 From the base circle, mark lengths equal to the lengths of the Y ordinates from the graph, and at each position draw a 20 mm diameter circle for the roller follower.
6 Draw the best profile for the cam so that the cam touches the rollers tangentially, as in the last example.

Case 5 (Fig. 24.10)

Cam specification:
Face cam, rotating clockwise. 12 mm diameter roller follower. Least radius of cam, 26 mm. Camshaft diameter, 30 mm.
0–180°, follower rises 30 mm with simple harmonic motion.
180–240°, dwell period.
240–360°, follower falls 30 mm with simple harmonic motion.

Fig. 24.10

1 Draw the cam graph, but note that for the first part of the motion the semi-circle is divided into six parts, and that for the second part it is divided into four parts.
2 Draw a base circle 32 mm radius, and divide into 30° intervals.
3 From each of the base-circle points, plot the lengths of the *Y* ordinates. Draw a circle at each point for the roller follower.
4 Draw a curve on the inside and the outside, tangentially touching the rollers, for the cam track.

The drawing shows the completed cam together with a section through the vertical centre line.

Note that the follower runs in a track in this example. In the previous cases, a spring or some resistance is required to keep the follower in contact with the cam profile at all times.

Case 6 (Fig. 24.11)

Cam specification:
Cylindrical cam, rotating anticlockwise, operating a roller follower 14 mm diameter. Cam cylinder. 60 mm diameter. Depth of groove, 7 mm.
0–180°, follower moves 70 mm to the right with simple harmonic motion.
180–360°, follower moves 70 mm to the left with simple harmonic motion.

1 Set out the cylinder blank and the end elevation as shown.

2 Divide the end elevation into 30° divisions.
3 Underneath the front elevation, draw a development of the cylindrical cam surface, and on this surface draw the cam graph.
4 Using the cam graph as the centre line for each position of the roller, draw 14 mm diameter circles as shown.
5 Draw the cam track with the sides tangential to the rollers.
6 Plot the track on the surface of the cylinder by projecting the sides of the track in the plan view up to the front elevation. Note that the projection lines for this operation do not come from the circles in the plan, except at each end of the track.
7 The dotted line in the end elevation indicates the depth of the track.
8 Plot the depth of the track in the front elevation from the end elevation, as shown. Join the plotted points to complete the front elevation.

Note that, although the roller shown is parallel, tapered rollers are often used, since points on the outside of the cylinder travel at a greater linear speed than points at the bottom of the groove, and a parallel roller follower tends to skid.

Fig. 24.11

Dimensioning cams

Fig. 24.12 shows a cam in contact with a roller follower; note that the point of contact between the cam and the roller is at A, on a line which joins the centres of the two arcs. To dimension a cam, the easiest method of presenting the data is in tabular form which relates the angular displacement $\varnothing$ of the cam with the radial displacement R of the follower. The cam could then be cut on a milling machine using these point settings. For accurate master cams, these settings may be required at half- or one-degree intervals.

Fig. 24.12

Spur gears

The characteristic feature of spur gears is that their axes are parallel. The gear teeth are positioned around the circumference of the pitch circles which are equivalent to the circumferences of the friction rollers in Fig. 24.13 below.

The teeth are of *involute* form, the involute being described as the locus traced by a point on a taut string as it unwinds from a circle, known as the base circle. For an involute rack, the base-circle radius is of infinite length, and the tooth flank is therefore straight.

Fig. 24.13

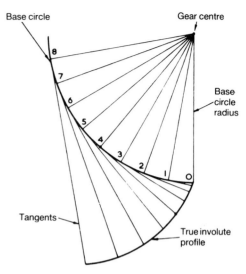

Fig. 24.14 Involute construction. The distance along the tangent from each point is equal to the distance around the circumference from point 0

The construction for the involute profile is shown in Fig. 24.14. The application of this profile to an engineering drawing of a gear tooth can be rather a tedious exercise, and approximate methods are used, as described later.

Spur-gear terms (Fig. 24.15)

The *gear ratio* is the ratio of the number of teeth in the gear to the number of teeth in the pinion, the pinion being the smaller of the two gears in mesh.

The *pitch-circle diameters* of a pair of gears are the diameters of cylinders co-axial with the gears which will roll together without slip. The pitch circles are imaginary friction discs, and they touch at the *pitch point*.

The *base circle* is the circle from which the involute is generated.

The *root diameter* is the diameter at the base of the tooth.

The *centre distance* is the sum of the pitch-circle radii of the two gears in mesh.

The *addendum* is the radial height of the tooth from the pitch circle to the tooth tip.

The *dedendum* is the radial depth of the tooth from the pitch circle to the root of the tooth.

The *clearance* is the algebraic difference between the addendum and the dedendum.

The *whole depth* of the tooth is the sum of the addendum and the dedendum.

The *circular pitch* is the distance from a point on one tooth to the corresponding point on the next tooth, measured round the pitch-circle circumference.

The *tooth width* is the length of arc from one side of the tooth to the other, measured round the pitch-circle circumference.

The *module* is the pitch-circle diameter divided by the number of teeth.

Fig. 24.15   Spur-gear terms

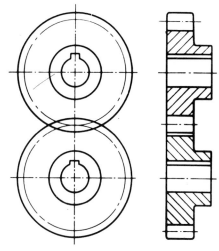

Fig. 24.16   Gear conventions

The *diametral pitch* is the reciprocal of the module, i.e. the number of teeth divided by the pitch-circle diameter.

The *line of action* is the common tangent to the base circles, and the *path of contact* is that part of the line of action where contact takes place between the teeth.

The *pressure angle* is the angle formed between the common tangent and the line of action.

The *fillet* is the rounded portion at the bottom of the tooth space.

The various terms are illustrated in Fig. 24.15.

Involute gear teeth proportions and relationships

$$\text{Module} = \frac{\text{pitch-circle diameter, PCD}}{\text{number of teeth, } T}$$

Circular pitch $= \pi \times$ module

$$\text{Tooth thickness} = \frac{\text{circular pitch}}{2}$$

Addendum = module
Clearance = 0.25 × module
Dedendum = addendum + clearance

Involute gears having the same pressure angle and module will mesh together. The British Standard recommendation for pressure angle is 20°.

The conventional representation of gears shown in Fig. 24.16 is limited to drawing the pitch circles and outside diameters in each case. In the sectional end elevation, a section through a tooth space is taken as indicated. This convention is common practice with other types of gears and worms.

Typical example using Professor Unwin's approximate construction

Gear data:
Pressure angle, 20°. Module, 12 mm: Number of teeth, 25.
Gear calculations:
Pitch-circle diameter = module × no. of teeth
$\qquad\qquad\qquad = 12 \times 25 = 300$ mm
Addendum = module = 12 mm
Clearance = 0.25 × module
$\qquad\qquad\quad = 0.25 \times 12 = 3$ mm
Dedendum = addendum + clearance = 12 + 3 = 15 mm
Circular pitch = $\pi \times$ module = $\pi \times 12$ = 37.68 mm
Tooth thickness = $\frac{1}{2} \times$ circular pitch = 18.84 mm

Stage 1 (Fig. 24.17)

a) Draw the pitch circle and the common tangent.
b) Mark out the pressure angle and the normal to the line of action.
c) Draw the base circle. Note that the length of the normal is the base-circle radius.

Stage 2 (Fig. 24.18)

a) Draw the addendum and dedendum circles. Both addendum and dedendum are measured radially from the pitch circle.
b) Mark out point A on the addendum circle and point B on the dedendum circles. Divide AB into three parts so that CB = 2AC.
c) Draw the tangent CD to the base circle. D is the point of tangency. Divide CD into four parts so that CE = 3DE.
d) Draw a circle with centre O and radius OE. Use this

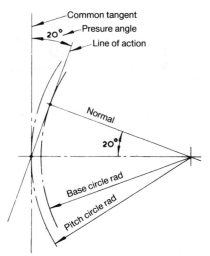

Fig. 24.17 Unwin's construction—stage 1

Fig. 24.18 Unwin's construction—stage 2

circle for centres of arcs of radius EC for the flanks of the teeth after marking out the tooth widths and spaces around the pitch-circle circumference.

Note that it may be more convenient to establish the length of the radius CE by drawing this part of the construction further round the pitch circle, in a vacant space, if the flank of one tooth, i.e. the pitch point, is required to lie on the line AO.

The construction is repeated in Fig. 24.19 to illustrate an application with a rack and pinion. The pitch line of the rack touches the pitch circle of the gear, and the values of the addendum and dedendum for the rack are the same as those for the meshing gear.

If it is required to use the involute profile instead of the approximate construction, then the involute must be constructed from the base circle as shown in Fig. 24.14. Complete stage 1 and stage 2(a) as already described, and mark off the tooth widths around the

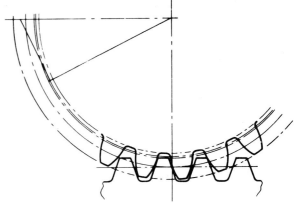

Fig. 24.19 Unwin's construction applied to a rack and pinion

pitch circle, commencing at the pitch point. Take a tracing of the involute in soft pencil on transparent tracing paper, together with part of the base circle in order to get the profile correctly oriented on the required drawing. Using a French curve, mark the profile in pencil on either side of the tracing paper, so that, whichever side is used, a pencil impression can be obtained. With care, the profile can now be traced onto the required layout, lining up the base circle and ensuring that the profile of the tooth flank passes through the tooth widths previously marked out on the pitch circle. The flanks of each tooth will be traced from either side of the drawing paper. Finish off each tooth by adding the root radius.

Helical gears

Helical gears have been developed from spur gears, and their teeth lie at an angle to the axis of the shaft. Contact between the teeth in mesh acts along the diagonal face flanks in a progressive manner; at no time is the full length of any one tooth completely engaged. Before contact ceases between one pair of teeth, engagement commences between the following pair. Engagement is therefore continuous, and this fact results in a reduction of the shock which occurs when straight teeth operate under heavy loads. Helical teeth give a smooth, quiet action under heavy loads; backlash is considerably reduced; and, due to the increase in length of the tooth, for the same thickness of gear wheel, the tooth strength is improved.

Fig. 24.20 illustrates the lead and helix angle applied to a helical gear. For single helical gears, the helix angle is generally 12°–20°.

Since the teeth lie at an angle, a side or end thrust occurs when two gears are engaged, and this tends to separate the gears. Fig. 24.21 shows two gears on parallel shafts and the position of suitable thrust bearings. Note that the position of the thrust bearings varies with the direction of shaft rotation and the 'hand' of the helix.

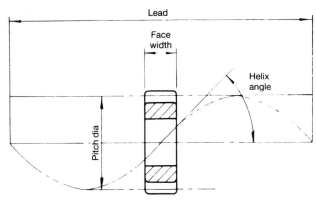

Fig. 24.20 Lead and helix angle for a helical gear

Fig. 24.21

In order to eliminate the serious effect of end thrust, pairs of gears may be arranged as shown in Fig. 24.22 where a double helical gear utilises a left- and a right-hand helix. Instead of using two gears, the two helices may be cut on the same gear blank.

Where shafts lie parallel to each other, the helix angle is generally 15°–30°. Note that a right-hand helix engages with a left-hand helix, and the hand of the helix must be correctly stated on the drawing. On both gears the helix angle will be the same.

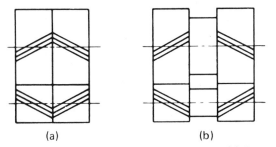

Fig. 24.22 Double helical gears. (a) On same wheel (b) On separate wheels.

For shafts lying at 90° to each other, both gears will have the same hand of helix, Fig. 24.23.

Helical gears can be used for shafts which lie at an

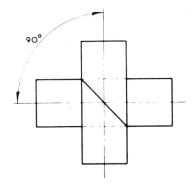

Fig. 24.23

angle less than 90°, but the hand of helix should be checked with a specialist gear manufacturer. The hand of a helix depends on the helix angle used and the shaft angles required.

Bevel gears

If the action of spur and helical gears can be related to that of rolling cylinders, then the action of bevel gears can be compared to a friction cone drive. Bevel gears are used to connect shafts which lie in the same plane and whose axes intersect. The size of the tooth decreases as it passes from the back edge towards the apex of the pitch cone, hence the cross-section varies along the whole length of the tooth. When viewed on the curved surface which forms part of the back cone, the teeth normally have the same profiles as spur gears. The addendum and dedendum have the same proportions as a spur gear, being measured radially from the pitch circle, parallel to the pitch-cone generator.

Data relating to bevel gear teeth is shown in Fig. 24.24. Note that the crown wheel is a bevel gear where the pitch angle is 90°. Mitre gears are bevel gears where the pitch-cone angle is 45°.

The teeth on a bevel gear may be produced in several different ways, e.g. straight, spiral, helical, or spiraloid. The advantages of spiral bevels over straight bevels lies in quieter running at high speed and greater load-carrying capacity.

The angle between the shafts is generally a right angle, but may be greater or less than 90°, as shown in Fig. 24.25.

Bevel gearing is used extensively in the automotive industry for the differential gearing connecting the drive shaft to the back axle of motor vehicles.

(a)

(b)

Bevel-gear terms and definitions

The following are additions to those terms used for spur gears.

The pitch angle is the angle between the axis of the gear and the cone generating line.

The root-cone angle is the angle between the gear axis and the root generating line.

The face angle is the angle between the tips of the teeth and the axis of the gear.

The addendum angle is the angle between the top of the teeth and the pitch-cone generator.

The dedendum angle is the angle between the bottom of the teeth and the pitch-cone generator.

The outside diameter is the diameter measured over the tips of the teeth.

The following figures show the various stages in drawing bevel gears. The approximate construction for the profile of the teeth has been described in the section relating to spur gears.

Gear data: 15 teeth, 20° pitch-cone angle, 100 mm pitch-circle diameter, 20° pressure angle.

Stage 1   Set out the cone as shown in Fig. 24.26.
Stage 2   Set out the addendum and dedendum. Project part of the auxiliary view to draw the teeth (Fig. 24.27)
Stage 3   Project widths A. B, and C on the outside, pitch, and root diameters, in plan view.
Complete the front elevation (Fig. 24.28).

A = addendum
B = dedendum
C = cone distance
D = pitch diameter
Da = outside diameter
Dr = root diameter
E = bottom clearance
F = face width
α (alpha) = addendum angle
β (beta) = dedendum angle
γ (gamma) = pitch angle
γa = back cone angle
γr = root angle
Σ (sigma) = shaft angle

Fig. 24.24   Bevel-gear terms
(a) Bevel gears
(b) Crown wheel and pinions

Shaft angle

Shaft angle

Fig. 24.25   Bevel-gear cones

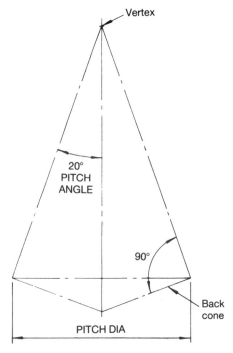

Vertex

20° PITCH ANGLE

90°

Back cone

PITCH DIA

Fig. 24.26   Stage 1

Fig. 24.27 Stage 2

Fig. 24.28 Stage 3

Worm gearing

Worm gearing is used to transmit power between two non-intersecting shafts whose axes lie at right angles to each other. The drive of a worm gear is basically a screw, or worm, which may have a single- or multi-start thread, and this engages with the wheel. A single-start worm in one revolution will rotate the worm wheel one tooth and space, i.e. one pitch. The velocity ratio is high; for example, a 40 tooth wheel with a single-start worm will have a velocity ratio of 40, and in mesh with a two-start thread the same wheel would give a velocity ratio of 20.

A worm-wheel with a single-start thread is shown in Fig. 24.29. The lead angle of a single-start worm is low, and the worm is relatively inefficient, but there is little tendency for the wheel to drive the worm. To transmit power, multi-start thread forms are used. High mechanical advantages are obtained by the use of worm-gear drives.

Worm-gear drives have many applications, for example indexing mechanisms on machine tools, torque converters, and slow-motion drives.

Fig. 24.30 shows typical cross-sections through part of a worm and wheel. Note the contour of the wheel, which is designed to give greater contact with the worm.

Recommendations for the representation of many types of gear assembly in sectional and simplified form are given in BS 308, Part 1.

Fig. 24.29

Fig. 24.30 Worm-gearing terms applied to a worm and part of a worm-wheel

Chapter 25
Springs

Mechanical springs may be defined as elastic bodies the primary function of which is to deform under a load and return to their original shape when the load is removed. In practice, the vast majority of springs are made of metal, and of these the greatest proportion are of plain-carbon steel.

Plain-carbon steels

These steels have a carbon-content ranging from about 0.5% to 1.1%. In general it may be taken that, the higher the carbon-content, the better the spring properties that may be obtained.

In the manufacture of flat springs and the heavier coil springs, it is usual to form the spring from annealed material and subsequently to heat-treat it. However, it is sometimes possible to manufacture flat springs from material which is already in the hardened and tempered condition, and this latter technique may give a lower production cost than the former.

For light coil springs, the material loosely known as piano wire is used; this is a spring wire which obtains its physical properties from cold-working, and not from heat-treatment. Springs made from this wire require only a low-temperature stress-relieving treatment after manufacture. Occasionally wire known as 'oil-tempered' is used—this is a wire which is hardened and tempered in the coil form, and again requires only a low-temperature stress relief after forming.

Plain-carbon steel springs are satisfactory in operation up to a temperature of about 180°C. Above this temperature they are very liable to take a permanent set, and alternative materials should be used.

Alloy steels

Alloy steels are essentially plain-carbon steels to which small percentages of alloying elements such as chromium and vanadium have been added. The effect of these additional elements is to modify considerably the steels' properties and to make them more suitable for specific applications than are the plain-carbon steels. The two widely used alloy steels are

a) chrome–vanadium steel—this steel has less tendency to set than the plain-carbon steels;

b) silicon–manganese steel—a cheaper and rather more easily available material than chrome-vanadium steel, though the physical properties of both steels are almost equivalent.

Stainless steels

Where high resistance to corrosion is required, one of the stainless steels should be specified. These fall into two categories.

a) *Martensitic*. These steels are mainly used for flat springs with sharp bends. They are formed in the soft condition and then heat-treated.

b) *Austenitic*. These steels are cold-worked for the manufacture of coil springs and flat springs, with no severe forming operations.

Both materials are used in service up to about 235°C.

High-nickel alloys

Alloys of this type find their greatest applications in high-temperature operation. The two most widely used alloys are

a) Inconel—a nickel–chrome–iron alloy for use up to about 320°C.

b) Nimonic 90—a nickel–chrome–cobalt alloy for service up to about 400°C, or at reduced stresses to about 590°C.

Both of these materials are highly resistant to corrosion.

Copper-base alloys

With their high copper-content, these materials have good electrical conductivity and resistance to corrosion. These properties make them very suitable for such purposes as switch-blades in electrical equipment.

a) *Brass*—an alloy containing zinc, roughly 30%, and the remainder copper. A cold-worked material obtainable in both wire and strip form, and which is suitable only for lightly stressed springs.

b) *Phosphor bronze*—the most widely used copper-base spring material, with applications the same as those of brass, though considerably higher stresses may be used.

c) *Beryllium copper*—this alloy responds to a precipitation heat-treatment, and is suitable for springs which contain sharp bends. Working stresses may be higher than those used for phosphor bronze and brass.

Compression springs

Fig. 25.1 shows two alternative types of compression springs for which drawing conventions are used. Note that the convention in each case is to draw the first and last two turns of the spring and to then link the space in between with a thin-chain line. The simplified representation shows the coils of the springs drawn as single lines.

Note. If a rectangular section compression spring is required to be drawn then the appropriate shape will appear in view (e), view (d) will be modified with square corners and the ∅ symbol in view (f) replaced by □.

A schematic drawing of a helical spring is shown in Fig. 25.2. This type of illustration can be used as a working drawing in order to save draughting time, with the appropriate dimensions and details added.

Fig. 25.3 shows four of the most popular end formations used on compression springs. When possible, grinding should be avoided, as it considerably increases spring costs.

Fig. 25.4 shows a selection of compression springs, including valve springs for diesel engines and injection pumps.

Flat springs

Fig. 25.5 shows a selection of flat springs, circlips, and spring pressings. It will be apparent from the selection that it would be difficult, if not impossible, to devise a drawing standard to cover this type of spring, and at present none exists.

Flat springs are usually made from high-carbon steel in the annealed condition, and are subsequently heat-treated; but the simpler types without bends can frequently be made more economically from material pre-hardened and tempered to the finished hardness required. Stainless steels are used for springs where considerable forming has to be done. For improved corrosion-resistance, 18/8 stainless steel is recommended; but, since its spring temper is obtained only by cold-rolling, severe bends are impossible. Similar considerations apply to phosphor bronze, titanium, and brass, which are hardened by cold-rolling. Beryllium copper, being thermally hardenable, is a useful material as it can be readily formed in the solution-annealed state.

(a) Conical compression springs with ground ends   (d) Cylindrical compression spring with ground ends

(b) Section convention   (e) Section convention

(c) Simplified representation   (f) Simplified representation

Fig. 25.1

Fig. 25.2   Schematic drawing of helical spring

Fig. 25.6 shows a selection of flat spiral springs, frequently used for brush mechanisms, and also for clocks and motors. The spring consists of a strip of steel spirally wound and capable of storing energy in the form of torque.

The standard for spiral springs is illustrated in Fig. 25.7, (a) and (b) show how the spring is represented in conventional and simplified forms.

If the spring is close wound and fitted in a housing then the illustrations in (c) and (d) are applicable.

(a) Closed ends, ground

(b) Closed ends

(c) Open ends, ground

(d) Open ends

Fig. 25.3

Fig. 25.4

Fig. 25.5

Fig. 25.6

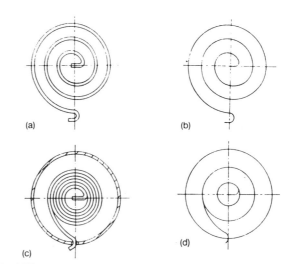

(a)

(b)

(c)

(d)

Fig. 25.7

Torsion springs

Various forms of single and double torsion springs are illustrated in Fig. 25.8.

Fig. 25.8

Fig. 25.9 gives a schematic diagram for a torsion spring. This type of drawing, adequately dimensioned, can be used for detailing.

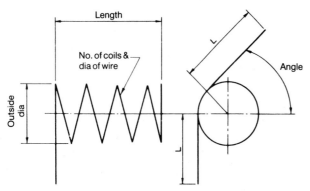

Fig. 25.9

The drawing conventions for a cylindrical right-hand helical torsion spring are shown in Fig. 25.10. (a) shows the usual drawing convention, (b) how to show the spring in a section and (c) gives the simplified representation.

Although torsion springs are available in many different forms, this is the only type to be represented in engineering-drawing standards. Torsion springs may be wound from square-, rectangular-, or round-section bar. They are used to exert a pressure in a circular arc, for example in a spring hinge and in door locks. The ends of the wire in the spring may be straight, curved, or kinked.

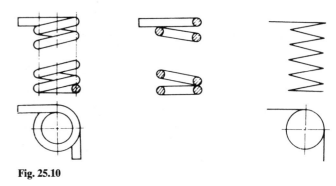

Fig. 25.10

Leaf springs

The two standards applicable to leaf springs are shown in Fig. 25.11. These springs are essentially strips of flat metal formed in an elliptical arc and suitably tempered. They absorb and release energy, and are commonly found applied to suspension systems.

(a)   (b)

(c)   (d)

Fig. 25.11
(a) and (b) conventional and simplified representations for a semi-elliptic leaf spring
(c) and (d) conventional and simplified representations for a semi-elliptic leaf spring with fixing eyes

Helical extension springs

A helical extension spring is a spring which offers resistance to extension. Almost invariably they are made from circular-section wire, and a typical selection is illustrated in Fig. 25.12.

The conventional representations of tension springs are shown in Fig. 25.13 and a schematic drawing for detailing is shown in Fig. 25.14.

Coils of extension springs differ from those of compression springs in so far as they are wound so close together that a force is required to pull them apart. A

Fig. 25.12

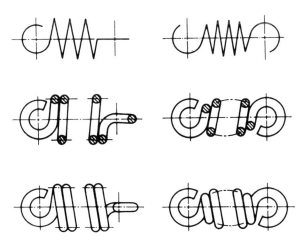

Fig. 25.13 Conventional representation of tension springs

Full eye

Half eye

Extended reduced eye

Extended full hook

Full double eye

V-hook

Coned end, reduced hook

Plain end, screw hook

Coned end, swivel bolt

Coned end, extended swivel hook

Square hook

Full hook

Fig. 25.15 Types of end loops

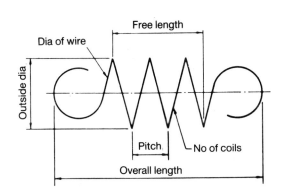

Fig. 25.14 Schematic drawing of tension spring

Outside dia

Dia of wire

Free length

Pitch

No of coils

Overall length

Fig. 25.16

variety of end loops is available for tension springs, and some of the types are illustrated in Fig. 25.15.

A common way of reducing the stress in the end loop is to reduce the moment acting by making the end loop smaller than the body of the spring, as shown in Fig. 25.16.

Table 25.1 Shows a selection of Belleville washers manufactured to DIN 6796 from spring steel to DIN 17222

FREE FROM BURRS

Notation	d_1 mm H 14	d_2 mm h 14	h mm max.[1]	min.[2]	s mm[3]	Force[4]	Test force[5]	Weight kg/1000 ≈	Core diameter
2	2.2	5	0.6	0.5	0.4	628	700	0.05	2
2.5	2.7	6	0.72	0.61	0.5	946	1100	0.09	2.5
3	3.2	7	0.85	0.72	0.6	1320	1500	0.14	3
3.5	3.7	8	1.06	0.92	0.8	2410	2700	0.25	3.5
4	4.3	9	1.3	1.12	1	3770	4000	0.38	4
5	5.3	11	1.55	1.35	1.2	5480	6550	0.69	5
6	6.4	14	2	1.7	1.5	8590	9250	1.43	6
7	7.4	17	2.3	2	1.75	11300	13600	2.53	7
8	8.4	18	2.6	2.24	2	14900	17000	3.13	8
10	10.5	23	3.2	2.8	2.5	22100	27100	6.45	10
12	13	29	3.95	3.43	3	34100	39500	12.4	12
14	15	35	4.65	4.04	3.5	46000	54000	21.6	14
16	17	39	5.25	4.58	4	59700	75000	30.4	16
18	19	42	5.8	5.08	4.5	74400	90500	38.9	18
20	21	45	6.4	5.6	5	93200	117000	48.8	20
22	23	49	7.05	6.15	5.5	113700	145000	63.5	22
24	25	56	7.75	6.77	6	131000	169000	92.9	24
27	28	60	8.35	7.3	6.5	154000	221000	113	27
30	31	70	9.2	8	7	172000	269000	170	30

[1] Max. size in delivered condition.
[2] Min. size after loading tests to DIN 6796.
[3] Permissable range of tolerance of s to DIN 1544 and DIN 1543 respectively for s > 6 mm.
[4] This force applies to the pressed flat condition and corresponds to twice the calculated value at a deflection $h_{mm} - s$.
[5] The test force applies for the loading tests to DIN 6796.

Disc springs

For bolted connections a very simple form of compression spring utilizes a hollow washer manufactured from spring steel, although other materials can be specified.

Table 25.1 shows a selection of Belleville washers manufactured to DIN 6796 from spring steel to DIN 17222.

If the disc has its top and bottom surfaces ground to approximately 95% of the appropriate thickness in the table above then bearing surfaces will be formed. These surfaces improve guidance where several discs are used together. Fig. 25.17 shows the disc spring with flats.

A disc spring stack with six springs in single series is given in Fig. 25.18. In this arrangement six times the number of individual spring deflections are available. The force available in the assembly is equivalent to that of a single disc spring. In single series concave and convex surfaces are opposed alternatively.

Fig. 25.19 shows three disc springs assembled in parallel with the convex surface nesting into the concave surface. Here the deflection available is equivalent to that of a single spring and the force equal to three times that of an individual disc.

The methods of assembly illustrated in Figs 25.18 and 25.19 can be combined to give many alternative selections of force and deflection characteristics. In the

0.95h

Fig. 25.17

Fig. 25.18

Fig. 25.19

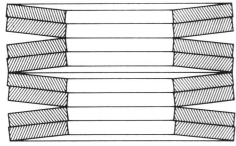

Fig. 25.20

stack given in Fig. 25.20 there are four disc spring components assembled in series and they each contain two disc springs assembled in parallel. This combination will give a force equal to twice that of an individual disc and a deflection of four times that of an individual disc.

Belleville washers are manufactured by Bauer Springs Ltd, of New Road, Studley, Warwickshire, B80 7LY where full specifications are freely available.

Drawing conventions for these springs are given in Fig. 25.21, and show (a) the normal outside view, (b) the view in section and (c) the simplified representation. These conventions can be adapted to suit the disc combination selected by the designer.

Fig. 25.21

Spring specifications

A frequent cause of confusion between the spring supplier and the spring user is lack of precision in specifying the spring. This often results in high costs due to the manufacturer taking considerable trouble to meet dimensions which are not necessary for the correct functioning of the spring.

It is recommended that, while all relevant data regarding the design should be entered on the spring detail drawing, some indication should be given as to which of the particular design points must be adhered

to for the satisfactory operation of the component; this is necessary to allow for variations in wire diameter and elastic modulus of the material. For example, a compression spring of a given diameter may, if the number of coils is specified, have a spring rate which varies between ± 15% of the calculated value. In view of this, it is desirable to leave at least one variable for adjustment by the manufacturer, and the common convenient variable for general use is the number of coils.

A method of spring specification which has worked well in practice is to insert a table of design data, such as that shown below, on the drawing. All design data is entered, and the items needed for the correct functioning of the spring are marked with an asterisk. With this method the manufacturer is permitted to vary any unmarked items, as only the asterisked data is checked by the spring user's inspector. The following are specifications typical for compression, tension, and torsion springs.

Compression spring

Total turns	7
Active turns	5
Wire diameter	1 mm
*Free length	12.7 ± 0.4 mm
*Solid length	7 mm max.
*Outside coil diameter	7.6 mm max.
*Inside coil diameter	5 mm
Rate	7850 N/m
*Load at 9 mm	31 ± 4.5 N
Solid stress	881 N/mm^2
*Ends	Closed and ground
Wound	Right-hand or left-hand
*Material	S202
*Protective treatment	Cadmium-plate

Tension spring

Mean diameter	11.5 mm
*O.D. max.	13.5 mm
*Free length	54 ± 0.5 mm
Total coils on barrel	$16\frac{1}{2}$
Wire diameter	1.62 mm
*Loops	Full eye, in line with each other and central with barrel of spring
Initial tension	None
Rate	2697 N/m
*Load	53 ± 4.5 N
*At loaded length	73 mm
Stress at 53 N	438 N/mm^2
Wound	Right-hand or left-hand
*Material	BS 1408 B
*Protective treatment	Lanolin

Torsion spring

Total turns on barrel	4
Wire diameter	2.6 mm
*Wound	Left-hand close coils
Mean diameter	12.7 mm
*To work on	9.5 mm dia bar
*Length of legs	28 mm
*Load applied at	25.4 mm from centre of spring

*Load   41 ± 2 N
*Deflection   20°
Stress at 41 N   595 N/mm²
*Both legs straight and
tangential to barrel
*Material   BS 1408 D
*Protective treatment   Grease

Wire forms

Many components are assembled by the use of wire forms which are manufactured from piano-type wire. Fig. 25.22 shows a selection, though the number of variations is limitless.

Fig. 25.22   Wire forms

Corrosion prevention

Springs operating under severe corrosive conditions are frequently made from phosphor bronze and stainless steel, and occasionally from nickel alloys and titanium alloys. For less severe conditions, cadmium- or zinc-plating is recommended; alternatively, there are other electroplatings available, such as copper, nickel, and tin. Phosphate coatings are frequently specified. Organic coatings, originally confined to stove enamels, now include many plastics materials such as nylon and polythene, as well as many types of epoxy resins.

Fatigue conditions

Many springs, such as valve springs, oscillate rapidly through a considerable range of stress and are consequently operating under conditions requiring a long fatigue life. The suitability of a spring under such conditions must be considered at the drawing-board stage, and a satisfactory design and material must be specified. Special treatments such as shot-peening or liquid-honing may be desirable. In the process of shot-peening, the spring is subjected to bombardment by small particles of steel shot; this has the effect of work-hardening the surface. Liquid honing consists of placing the spring in a jet of fine abrasive suspended in water. This has a similar effect to shot-peening, and the additional advantage that the abrasive stream removes or reduces stress raisers on the spring surface.

Chapter 26
Welding and welding symbols

In general, welding may be described as a process of uniting two pieces of metal or alloy by raising the temperature of the surfaces to be joined so that they become plastic or molten. This may be done with or without the application of pressure and with or without the use of added metal. This definition excludes the more recently developed method of *cold-welding*, in which pressure alone is used. Cold-welding, however, has a limited application, and is used principally for aluminium and its alloys, and not for steel.

There are numerous methods of welding, but they can be grouped broadly into two categories. *Forge welding* is the term covering a group of welding processes in which the parts to be joined are heated to a plastic condition in a forge or other furnace, and are welded together by applying pressure or impact, e.g. by rolling, pressing, or hammering. *Fusion welding* is the process where the surfaces to be joined are melted with or without the addition of filler metal. The term is generally reserved for those processes in which welding is achieved by fusion alone, without pressure.

Forge welding will be dealt with first. *Pressure welding* is the welding of metal by means of mechanical pressure whilst the surfaces to be joined are maintained in a plastic state. The heating for this process is usually provided by the process of *resistance welding*, where the pieces of metal to be joined are pressed together and a heavy current is passed through them.

Projection welding is a resistance-welding process in which fusion is produced by the heat obtained from the resistance to flow of electric current through the work parts, which are held together under pressure by the electrodes providing the current. The resulting welds are localized at predetermined points by the design of the parts to be welded. The localization is usually accomplished by projections or intersections.

Spot welding is a resistance-welding process of joining two or more overlapping parts by local fusion of a small area or 'spot'. Two copper-alloy electrodes contact either side of the overlapped sheets, under known loads produced by springs or air pressure. *Stitch welding* is spot welding in which successive welds overlap. *Seam welding* is a resistance-welding process in which the electrodes are discs. Current is switched on and off regularly as the rims of the discs roll over the work, with the result that a series of spot welds is produced at such points. If a gas-tight weld is required, the disc speed and time cycle are adjusted to obtain a series of overlapping welds.

Flash-butt welding is a resistance-welding process which may be applied to rod, bar, tube, strip, or sheet to produce a butt joint. After the current is turned on, the two parts are brought together at a predetermined rate so that discontinuous arcing occurs between the two parts to be joined. This arcing produces a violent expulsion of small particles of metal (flashing), and a positive pressure in the weld area will exclude air and minimise oxidation. When sufficient heat has been developed by flashing, the parts are brought together under heavy pressure so that all fused and oxidised material is extruded from the weld.

Fusion-welding processes can now be dealt with. The heat for fusion welding is provided by either gas or electricity. *Gas welding* is a process in which heat for welding is obtained from a gas or gases burning at a sufficiently high temperature produced by an admixture of oxygen. Examples of the gases used are acetylene (oxy-acetylene welding), hydrogen (oxy-hydrogen welding), and propane (oxy-propane welding). In *air-acetylene welding*, the oxygen is derived from the atmosphere by induction.

Electrical fusion welding is usually done by the process of 'arc welding'. *Metal-arc welding* is welding with a metal electrode, the melting of which provides the filler metal. *Carbon-arc welding* is a process of arc welding with a carbon electrode (or electrodes), in which filler metal and sometimes flux may be used. *Submerged-arc welding* is a method in which a bare copper-plated steel electrode is used. The arc is entirely submerged under a separate loose flux powder which is continually fed into and over the groove which is machined where the edges to be welded are placed together. Some of the flux powder reacts with the molten metal: part fuses and forms a refining slag which solidifies on top of the weld deposit; the remainder of the powder covers the weld and slag, shielding them from atmospheric contamination and retarding the rate of cooling.

Argon-arc welding is a process where an arc is struck between an electrode (usually tungsten) and the work in an inert atmosphere provided by directing argon into the weld area through a sheath surrounding the electrode. *Heliarc welding* uses helium to provide the inert atmosphere, but this process is not used in the United Kingdom, because of the non-availability of helium. Several proprietary names are used for welding processes of this type, e.g. *Sigma (shielded inert-gas metal-arc) welding* uses a consumable electrode in an argon atmosphere. *Atomic-hydrogen arc welding* is a process where an alternating-current arc is maintained between tungsten electrodes, and each electrode is surrounded by an annular stream of hydrogen. In passing through the arc, the molecular hydrogen is dissociated into its atomic state. The recombination of the hydrogen atoms results in a very great liberation of heat which is used for fusing together the metals to be joined. *Stud welding* is a process in which an arc is struck between the bottom of a stud and the base metal. When a pool of molten metal has formed, the arc is extinguished and the stud is driven into the pool to form a weld.

The application of welding symbols to working drawings

The notes following are meant as a guide to the method of applying the more commonly used welding symbols relating to the simpler types of welded joints to engineering drawings. BS 499: Part 2 gives the complete scheme. The standard is not intended to apply to complex joints involving multiple welds since it may be easier to detail such joints and welds on a separate drawing.

Each type of weld is characterised by a symbol given in Table 26.1. Note that the symbol is representative of the shape of the weld, or the edge preparation, but does not indicate any particular welding process and does not specify either the number of runs to be deposited or whether or not a root gap or backing material is to be used. These details would be provided on a welding procedure schedule for the particular job.

It may be necessary to specify the shape of the weld surface on the drawing as flat, convex or concave and a supplementary symbol, shown in Table 26.2, is then added to the elementary symbol. An example of each type of weld surface application is given in Table 26.3.

A joint may also be made with one type of weld on a particular surface and another type of weld on the back and in this case elementary symbols representing each type of weld used are added together. The last example in Table 26.3 shows a single-V butt weld with a backing run where both surfaces are required to have a flat finish.

A welding symbol is applied to a drawing by using a reference line and an arrow line as shown in Fig. 26.1. The reference line should be drawn parallel to the bottom edge of the drawing sheet and the arrow line forms an angle with the reference line. The side of the joint nearer the arrow head is known as the 'arrow side' and the remote side as the 'other side'.

The welding symbol should be positioned on the reference line as indicated in Table 26.4.

Sketch (a) shows the symbol for a single-V butt weld below the reference line because the external surface of the weld is on the arrow side of the joint.

Table 26.1   Elementary weld symbols

Form of weld	Illustration	BS symbol
Butt weld between flanged plates (the flanges being melted down completely)		
Square butt weld		
Single-V butt weld		
Single-bevel butt weld		
Single-V butt weld with broad root face		
Single-bevel butt weld with broad root face		
Single-U butt weld		
Single-J butt weld		
Backing or sealing run		
Fillet weld		
Plug weld (circular or elongated hole, completely filled)		
Spot weld (resistance or arc welding) or projection weld	(a) Resistance / (b) Arc	
Seam weld		

Table 26.2   Supplementary symbols

Shape of weld surface	BS symbol
flat (usually finished flush)	——
convex	⌒
concave	⌣

Table 26.3   Some examples of the application of supplementary symbols

Form of weld	Illustration	BS symbol
Flat (flush) single-V butt weld		▽
Convex double-V butt weld		⋈
Concave fillet weld		◿
Flat (flush) single-V butt weld with flat (flush) backing run		⊻

Table 26.4   Significance of the arrow and the position of the weld symbol

Illustration	Graphic representation		Symbolic representation	
(a)				
(b)				
(c)				
(d)				

Sketch (b) shows the same symbol above the reference line because the external surface of the weld is on the other side of the joint.

Sketch (c) shows the symbol applied to a double-V butt weld.

Sketch (d) shows fillet welds on a cruciform joint where the top weld is on the arrow side and the bottom weld is on the other side.

The positioning of the symbol is the same for drawings in first or third angle projection.

Additional symbols can be added to the reference line as shown in Fig. 26.2. Welding can be done in the

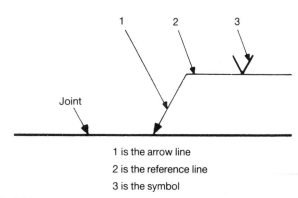

1 is the arrow line
2 is the reference line
3 is the symbol

Fig. 26.1

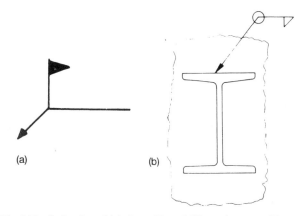

Fig. 26.2   Indication of (a) site welds and (b) continuous welds

factory or on site when plant is erected. A site weld is indicated by a flag. A continuous weld all round a joint is shown by a circle at the intersection of the arrow and the reference line. Note that if a continuous weld is to be undertaken at site then both symbols should be added to the drawing.

If it is necessary to indicate on a drawing that non-destructive testing of a joint is to be undertaken then the following symbol should be added near to the end of the reference line.

The introductory notes relating to welding processes are of a general nature. There are many specialized methods listed in Table 10, BS 499; Part 2. Each process is given an individual identification number and group headings are as follows; a) Arc welding, b) Resistance welding, c) Gas welding, d) Solid phase welding; Pressure welding, e) Other welding processes, f) Brazing, soldering and braze welding.

The table complies with International Standard ISO 4063.

A welding procedure sheet will usually give details of the actual process to be used on a particular joint. On the drawing, a reference line with an arrow pointing towards the joint at one end, will have a 'fork' added at the other containing the selected number. In the example given below, the figure 23 indicates that projection welding is the chosen method.

Useful standards for the draughtsman are as follows;
BS 499; Part 1; Glossary for welding, brazing and thermal cutting.
BS 499; Part 2; Specification for symbols for welding.
BS 499; Part 2C; Welding symbols. Provides examples of welding symbols applications in chart form.

Dimensions of welds

The dimensions of a weld may be added to a drawing in the following manner.

Dimensions relating to the cross section of the weld are symbolised by *b* and are given on the left-hand side of the symbol. The cross-sectional dimension to be indicated for a fillet weld is the leg length. If the design throat thickness is to be indicated then the leg-length dimension is prefixed with the letter b and the design throat thickness with the letter a.

Longitudinal dimensions are symbolized by *l* and are given on the right-hand side of the symbol. If the weld is not continuous then distances between adjacent weld elements are indicated in parentheses. Unless dimensional indication is added to the contrary, a fillet weld is assumed to be continuous along the entire length of the weld. Leg-length dimensions of fillet welds of 3,4,5,6,8,10,12,16,18,20,22 and 25 millimetres are the preferred sizes.

Applications of dimensions to different types of fillet welds are given in Table 26.5 in order to indicate the scope of the British Standard, which should be consulted to fully appreciate this topic. Table 26.5(a) shows dimensions applied to continuous fillet welds, (b) shows dimensions applied to intermittent fillet welds and (c) shows dimensions applied to staggered intermittent fillet welds.

Table 26.5   The dimensioning of welds

Definition	Inscription
(a)	

b is the leg length
a is the design throat thickness

| (b) | |

l is the length of weld (without end craters)
e is the distance between adjacent weld elements
n is the number of weld elements
b and *a* as defined above

| (c) | |

a, b, l, e and *n* as defined above

Chapter 27
Engineering diagrams

The following list gives details of Standards and Publications which will provide a major source of reference material for use in the preparation of various types of Engineering diagram.

BS 1192 is published in 5 parts and covers the practice of construction drawing. Details are as follows:

Part 1. Recommendations for general principles. Sets out general principles for the preparation of all drawings for the construction industry, including schedules prepared on drawing sheets and numbered as one of a drawing set. Applicable to drawings produced by computer techniques, and to drawings that will be reproduced by microfilming in accordance with BS 5536.

Part 2. Recommendations for architectural and engineering drawings. Illustrates the general principles given in *Part 1* and the application of the symbols and conventions given in *Part 3*.

Part 3. Recommendations for symbols and other graphic conventions.

Part 4. Recommendations for landscape drawings. Includes symbols and abbreviations which are used in a series of typical drawings.

Part 5. Guide for structuring of computer graphic information. Gives guidance on the representation of construction by computers, primarily for the purpose of generating drawings and the exchange of data between CAD users.

BS 1553: Specification for graphical symbols for general engineering.

Part 1. Piping systems and plant. This section deals with graphical symbols for use in the creation of flow and piping plant and heating and ventilation installations.

Part 2. Graphical symbols for power generating plant. Includes steam and internal combustion engines and turbines, also auxiliary plant.

Part 3. Graphical symbols for compressing plant. Includes applications to air operated switchgear.

BS 2917: Specification for graphical symbols used on diagrams for fluid power systems and components. This standard shows symbols to be used on diagrams of hydraulic and pneumatic transmission systems.

Graphical symbols for electrical power, telecommunications and electronics diagrams are covered by BS 3939. This field of technology is expanding at a rapid rate and the current situation is described in the foreword to BS 3939: Part 12: 1991 as follows;

This Part of BS 3939, prepared under the direction of the General Electrotechnical Standards Policy Committee, recommends graphical symbols, with definitions, for the representation of binary logic elements and functions, for use in electrical power, telecommunications and electronics diagrams. It is identical with the English text in IEC Publication 617–12: 1991 'Graphical symbols for diagrams', Part 12 'Binary logic elements', published by the International Electrotechnical Commission (IEC).

This Part of BS 3939 supersedes the previous edition which is withdrawn in its entirety.

The title of the other Parts in the BS 3939 series are as follows:

Part 1   General information, general index
Part 2   Symbol elements, qualifying symbols and other symbols having general application
Part 3   Conductors and connecting devices
Part 4   Passive components
Part 5   Semiconductors and electron tubes
Part 6   Production and conversion of electrical energy
Part 7   Switchgear, controlgear and protective devices
Part 8   Measuring instruments, lamps and signalling devices
Part 9   Telecommuncations: switching and peripheral equipment
Part 10 Telecommunications: transmission
Part 11 Architectural and topographical installation plans and diagrams
Part 13 Analogue elements

Titles to any further Parts will be added in due course.

To take account of the development of computer draughting, the symbols have been designed to a grid. A transparent over-lay with this grid is included in Part 1.

Cross-references

International standard	Corresponding British Standard
	BS 3939 Guide for graphical symbols for electrical power, telecommunications and electronics diagrams

IEC 617–2: 1983	Part 2: 1985 Symbol elements, qualifying symbols and other symbols having general application (Identical)
IEC 617–3: 1983	Part 3: 1985 Conductors and connecting devices (Identical)
IEC 617–10: 1983	Part 10: 1985 Telecommunications: transmission (Identical)
IEC 617–13: 1983	Part 13: 1985 Analogue elements (Identical)
ISO 31–11: 1978	BS 5775 Specification for quantities, units and symbols Part 11: 1979 Mathematical signs and symbols for use in the physical sciences and technology (Identical)

The Technical Committee has reviewed the provisions of IEC 113–7, to which reference is made in the text, and has decided that they are acceptable for use in conjunction with this standard.

A selection of symbols from BS 3939 is given in the Education Publication PP 7303: *Electrical and Electronic Graphical Symbols for Schools and Colleges.*

Engineering diagram drawing practice is covered by BS 5070 and is available in four parts,

Part 1. Recommendations for general principles.

Sets out drawing practice common to all fields on engineering diagrams. Covers layout, types of diagrams, drawing sheets, lines, lettering and graphical symbols.

Part 2. Recommendations for electrotechnology diagrams.

Gives principles and presentation for electrical, electronic and telecommunication diagrams. It includes block, circuit, wiring, interconnection and supplementary diagrams. To be used in conjunction with BS 3939 and BS 5070:Part 1.

Part 3. Recommendations for mechanical/fluid flow diagrams.

Gives principles and presentation for mechanical, hydraulic, pneumatic topographic, block, circuit, piping, interconnection and supplementary diagrams. To be used in conjunction with BS 5070:Part 1.

Part 4. Recommendations for logic diagrams.

Principles and presentation. Covers signal names, characteristics and logic circuit diagrams. To be read in conjunction with Parts 1 and 2 of BS 5070.

PP 7307. Graphical symbols for use in schools and colleges, is a reference manual of graphical symbols selected from a number of British Standards, together with examples of diagrams and background information.

The diagrams which follow are representative of various branches of engineering and obviously every application will be different. The examples can only indicate the type of diagram one is likely to encounter. The standards listed will provide a valuable source of information relating to layout, content and the appropriate symbols to be used.

Table 27.1

Note – The longer line represents the positive pole, the short line the negative pole. The short line may be thickened for emphasis

Table 27.1 *continued*

Inductor, coil, winding, choke

Inductor with magnetic core

Inductor with tappings, two shown

Transformer with magnetic core

Machine, general symbol
The asterisk is replaced
by a letter designation
as follows:
C   synchronous converter
G   generator
GS synchronous generator
M   motor
MG machine capable of use as
a generator or motor
MS synchronous motor

Socket (female), pole of a socket

Plug (male), pole of plug

Plug and jack, telephone type, two-pole
Note – The longest pole on the plug represents the tip of the plug, and the shortest the sleeve

Coaxial plug and socket

Variability, non-inherent

Pre-set adjustment

Resistor, general symbol

Variable resistor

Resistor with sliding contact

Heating element

Capacitor, general symbol

Polarized capacitor, for example, electrolytic

Variable capacitor

Make contact normally open, also general symbol for a switch

Break contact

Change-over contact, break before make

Contactor, normally open

Contactor, normally closed

Operating device (relay coil), general symbol

Coil of an alternating current relay

Mechanical coupling

Wattmeter

Watt-hour meter

Signal lamp, general symbol

Spotlight

Luminaire, fluorescent, general symbol

With three fluorescent tubes

Bell

Buzzer

Microphone

Earphone

Loudspeaker

Transducer head, general symbol

Clock, general symbol

Laser (optical maser, general symbol)

Amplifier

Semiconductor diode, general symbol

Light emitting diode, general symbol

Tunnel diode

Triode thyristor, type unspecified

PNP transistor

NPN transistor with collector connected to the envelope

Unijunction transistor with P-type base

Junction field effect transistor with N-type channel

Photovoltaic cell

Engineered systems

All of the engineering specialities referred to at the start of this chapter need diagrams and circuits in order to plan and organize the necessary work. It is very difficult to standardize aspects of work of such a varied nature, however, the following general notes are applicable in most circumstances.

Block diagrams

Block symbols or outlines are used to indicate the main separate elements in an installation and how they are functionally linked together. The diagram needs to be simple, so that the basics of the operation it represents, can be appreciated quickly.

The block symbols refer to single hardware items or self contained units without necessarily indicating all of the exact connections.

A block diagram can be presented to show a sequence of events to the reader and be used for fault diagnosis.

Circuit diagrams

The term circuit suggests electrical components wired together but this need not be the only case. The circuit could show parts of a central heating system connected by water piping or units in an air conditioning system joined together by fabricated ductwork.

Theoretical circuit diagrams

Design staff will prepare theoretical circuit diagrams where all the necessary connections for the correct operation of the system are included. Different sections of industry freely use other terms, such as schematics, sequence diagrams and flow charts.

In all these diagrams the component parts are arranged neatly and if possible horizontally or vertically. If several diagrams form a set, then the style of presentation should be consistent.

One of the conventions with this type of diagram is that components should be arranged so that the sequence of events can be read from left to right, or top to bottom, or perhaps a combination of both.

The diagram does not differentiate between the physical sizes of the separate components. The actual component shape may not be reflected in the standardized symbols and the arrangement on the diagram will not attempt to indicate the true layout of all the items.

Basic engineering practice follows where specifications will be produced for all parts of the system covering, for example, the components in detail, materials, manufacturing processes, relevant standards, inspection procedures, delivery dates and costs. The customer needs to know exactly what is being supplied, and details of financial arrangements. Contracts will be exchanged when supplier and client are satisfied. Obviously failure of any aspect of an agreement may involve either party in financial loss and litigation could follow. It is of course in nobody's interest that this should occur.

Construction diagrams

When the system is engineered, the actual position of each component part will dictate the arrangement of wiring, piping and general services, etc. The engineer will need to divide the work according to the scale of the contract and define which diagrams are necessary.

A production circuit diagram for an electrical control panel would show how the panel was built, with all the necessary line and neutral wiring connections in their exact places together with earth links. An exact construction record is essential for service requirements. Wiring must be sized. Standards for current capacity

Fig. 27.1

dictate the wire dimensions. Wires are often colour-coded to facilitate tracing. Wires are run singly, in multicored cables, in looms and conduits, inside and outside and in almost every conceivable ambient situation.

A selection of binary logic elements is also included in PP 7303 and an application for a one-stage accumulator shown in Fig. 27.1.

Note. PP 7303:1986 and associated wall chart.

PD 7303:1981 was revised to bring it into line with IEC 617:1983. The symbols are not fundamentally affected but have been shown with one thickness of line in most cases, often simplified. The changes should help when using the symbols in CAD.

The following electronic diagrams include symbols from BS 3939 and the extracts from BS 5070 are given by kind permission from British Standards. Note the clear, accurate and presentable layout which is essential in the production of engineering diagrams. Clarity depends on sufficient thought being given to spacing not only the symbols but associated notes and references.

Fig. 27.2 shows a thyristor control system.

Part (a) outlines the basic blocks.

Part (b) provides added details to the four parts.

Part (c) gives the component connections for the zero voltage trigger with waveforms at various points.

Part (d) is an example of a supplementary diagram where the waveforms are related to a common datum.

Note. This diagram forms an introduction to the basic system concept highlighting the major functional areas of the overall system. It is intended for users who require a basic appreciation of the system.

(a) Block diagram

Note. This diagram shows part of a detailed block diagram developed from the simple form of block diagram shown in (a). In this diagram functional information has been expanded and specific information in respect of input/output terminations has been added.

At this level the diagram becomes a useful diagnostic tool where input/output parameters may be monitored and hence faulty operation detected at unit, printed boards, etc. level. Maintenance at this level involves the replacement of the faulty unit or printed circuit board thus restoring normal working fairly rapidly.

(b) Detailed block diagram

Fig. 27.2

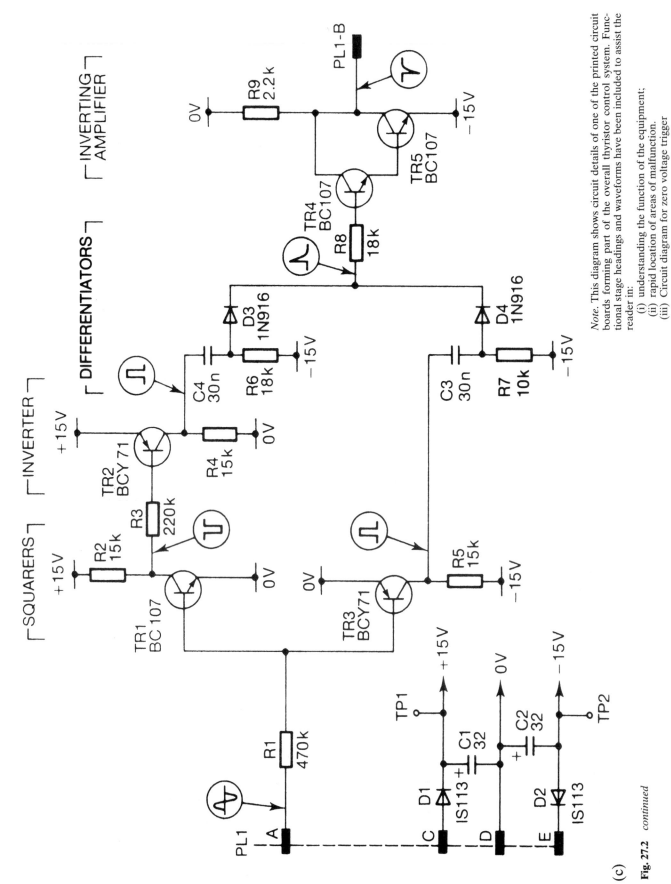

Note. This diagram shows circuit details of one of the printed circuit boards forming part of the overall thyristor control system. Functional stage headings and waveforms have been included to assist the reader in:

(i)   understanding the function of the equipment;
(ii) rapid location of areas of malfunction.
(iii) Circuit diagram for zero voltage trigger

(c)

Fig. 27.2 *continued*

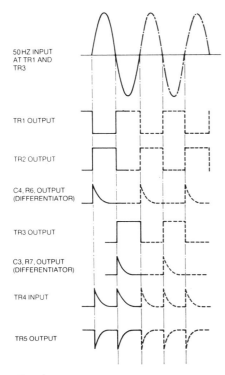

50 HZ INPUT
AT TR1 AND
TR3

TR1 OUTPUT

TR2 OUTPUT

C4, R6, OUTPUT
(DIFFERENTIATOR)

TR3 OUTPUT

C3, R7, OUTPUT
(DIFFERENTIATOR)

TR4 INPUT

TR5 OUTPUT

Fig. 27.2 *continued*

Note. Although the waveforms shown in (c) are of assistance to the reader in establishing that the correct signals are present at various parts of the diagram, thus assisting preliminary fault location, it does not show the timing relationship that exists between waveforms. This diagram is a supplementary diagram that defines this time relationship.

(d) Supplementary diagram for waveform development for zero-voltage trigger

Wiring diagrams for motor vehicles

The following diagrams are reproduced by kind permission of the Ford Motor Company Ltd and show part of the wiring circuit for the Transit van. Service manuals need to be presented so that the technician can easily check each function for satisfactory operation. The manual is therefore written with each circuit shown completely and independently in one chapter or 'cell'. Other components which are connected to the circuit may not be shown unless they influence the circuit operation. For the benefit of the user, the diagram needs to be reasonably large to be used while work proceeds.

Fig. 27.3 shows part of the circuitry for the headlamps. Each cell normally starts with the fuse, ignition switch, etc. that powers the circuit. Current flows from the power source at the top of the page towards earth at the bottom. Within the schematic diagram, all switches, sensors and relays are shown 'at rest' as if the ignition switch was in the OFF position.

Page numbering system The Ford Motor Company procedure is to divide the electrical system into individual sections. For example, the Engine Control Section is 29. The section is further broken down into cells, where Cell 29.00 is the 2.OL2V engine, Cell 29.10 is the 2.OLEFI engine and so on. All the engine information can be found in cells 29.00 to 29.50. Finally the pages in the manual are numbered using the cell number, a dash, and then a consecutive number. If there are two pages in Cell 29.00, they will be numbered 29.00–00 and 29.00–01.

The headlamp circuit is continued at points 'A' & 'B' on sheet number 32.00–02 down to the ground point G1001. The location of ground points is given on a separate diagram.

Connections to the Dim light relay, component K20, at the centre left side of diagram 32.00–01 are made via a connector, numbered C122. Connectors are grouped together on separate sheets and 91.00–05 (Fig. 27.5) is typical. The dim light relay connector is shown at the right centre side of this sheet. Looking at the connector, each wire is numbered to correspond with the circuit diagram, and recognizable by the colour coding.

Vehicle wiring is made up into looms and at the end of each of the bunches the connectors are fitted. Diagram 90.10–22 (Fig. 27.4) shows a pictorial view of wiring looms in the cab, and this is one of a series.

The circuits for Signal and Hazard indicators, also the Wiper/Washer control are given on diagrams 32.40–00 and 32.60–01 (Fig. 27.3). Note that the switch N9 serves three functions. On 32.00–01 its function is to control Side/tail and Dipped beam on either side of the vehicle. On 32.60–01, the four wipe functions are actuated. Note also that on 32.40–00 the multi-function switch is illustrated in full. However four connections on this switch relate to headlamp functions so these details are given in full from points 8, 9, 10 & 11 on 32.00–01. The location of connector C3 at position D1 on diagram 90.10–22 shows the connection to the Direction indicators at the top left corner on 32.40–00.

There are many pages similar to these in a vehicle service manual. This type of simplified layout has many advantages during servicing and fault finding operations.

(a) Headlamps 32.00–01 (Ford Motor Co. System).

Fig. 27.3

(b) Headlamps 32.00–02.

Fig. 27.3 *continued*

(c) Signal and hazard indicators 32.40–00.

Fig. 27.3 *continued*

(d) Wiper/washer control 32.60–01.

Fig. 27.3 *continued*

Component Location Views

Behind dash (LH drive shown, RH drive similar)

A2 ..... Instrument cluster	D 1
C2	E 0
C3	D 1
C6	D 1
C7	E 2
C8	C 4
C9	D 3
C39	A 4
C49	E 1
C68	B 4
C71	E 4
C93	C 4
C123	D 1
C162	E 1
C339	C 4
C340	D 4
C105a	A 3
C105b	F 4
E4 ..... Trailer warning light	E 1
G1001	E 4
M3 ..... Heater blower	E 2
N3 ..... Door ajar sensor, passenger side	F 4

N13 ... Electrical headlamp alignment	
.... system switch	E 1
N14 ... Heater blower switch	D 3
N15 ... Stop light switch	·D 1
N16 ... Safety switch	D 1
N26 ... Door ajar sensor, driver side	A 3
S1	B 4
S2	B 4
S3	C 4
S4	D 4
S5	C 4
S6	B 4
S7	B 4
S8	B 4
S10	B 4
S11	B 4
S12	E 4
S13	E 4
S18	B 4

Fig. 27.4 Component location views 90.10–22.

Connector Views

Fig. 27.5 Connector views 91.00–05.

Heating, ventilation and air conditioning systems

Control systems are devised to suit each individual application. Generally, each part of the system will contain air of different types. With reference to Fig. 27.6 the room air (RA) is extracted by a fan, a proportion of the air is exhausted to atmosphere and the remainder is returned and mixed with a supply of fresh outside air. The mixed air will then be returned to the room via a supply air fan after its temperature has been corrected to suit the design requirements.

In most cases this involves a heating operation. However, if the outside air temperature is high, or if there are considerable heat gains within the controlled space, then a cooling operation may well be necessary. In addition, full air conditioning specifications require control of the relative humidity in the space.

Personal physical comfort conditions depend on adjusting air and surface temperatures, humidity and air movement. By balancing these four factors, the engineer can design a climate to suit any type of activity.

In Fig. 27.6 the air is heated by passage through a heat exchanger supplied with hot water. Hot water from a boiler operating at normal atmospheric pressure is Low Temperature Hot Water (LTHW). If the boiler operates at pressure, its output is High Temperature Hot Water (HTHW). The heat exchanger could also be supplied with steam or operated by electricity.

The volume of outside air supplied will vary considerably with the occupancy density within the space and the activity. For example, theatres, public houses, conference rooms, areas with large solar heat gains, industrial premises with processing equipment, swimming pools and incubators, to name just a few, all require very special attention. Hence, various degrees of air purification and levels of sophistication exist. Three typical schemes now follow and Fig. 27.7 shows a ventilation system. Here a controller adjusts the position of a three way valve so that more, or less, water passes through the heat exchanger in response to supply air temperature demands. The air supplies are controlled by electrically operated dampers fitted in the ducts. Note that the air into the space has its temperature measured by the sensor f1. An alternative position for the sensor could be in the outlet duct when it could take note of any temperature increases gen-

Air types: (DIN 1946).

Air type	Abbreviation
Outside air	AU
Penetrated outside air	VAU
Exhaust air	FO
Aftertreated exhaust air	NFO
Extract air	AB
Room air	RA
Aftertreated extract air	NAB
Return air	UM
Mixed air	MI
Supply air	ZU
Pretreated supply air	VZU
Water:	
Flow	V
Return	R

erated within the space, or it could be positioned within the space itself, shown by u2. Many choices need to be considered.

A scheme for partial air conditioning is illustrated in Fig. 27.8 where in addition to ventilation and heating, the humidity has been given a degree of control. For full air conditioning it is necessary to provide equipment to cool the air and typical plant has been added to the layout in Fig. 27.9.

All of the installed plant needs to be carefully sized to ensure that specifications for air quality are met. The

Fig. 27.7 Ventilation system diagram. System designed to control the temperature of supply air into a space with heating from LTHW, HTHW or a steam heated coil. Variant with remote setpoint potentiometer.

Functions

Temperature control

The duct sensor f1 measures the temperature t_{zu}. The controller u1 compares this value with the selected setpoint X_K on the controller u1 or on the remote setpoint potentiometer u2 and adjusts the heating coil valve s1 in accordance with the difference between the two.

Safety devices

When there is danger of frost, the frost protection thermostat f2 must switch off the fan, close the damper s2, open the heating coil valve s1 and, where appropriate, switch on the heating pump.

Controller output temperature

Functions

Temperature control

The duct sensor f1 or the room sensor f2 measures the temperature t_R. The controller u1 compares this value with the selected setpoint X_K and adjusts the heating coil step controller (or power controller) u2 in accordance with the difference between the two.

Humidity control

The duct sensor f1 or the room sensor f2 measures the humidity φ_R. The controller u3 compares this value with the selected setpoint X_K and adjusts the humidfying valve s1 in accordance with the difference between the two.

Safety devices

If there is an air flow failure, the differential pressure switch f4 must cut off the control voltage of the electric heating coil. With electric heating coils, it is advisable to incorporate a timer so that the fan will run on for approx. 5 min and dissipate any residual heat.

Fig. 27.8 Partial air conditioning system. System designed to control the extract air from a room. The air into the space is heated with an electric heating coil and humidified with steam. **Alternative:** Room sensor instead of extract air sensor. **Variant:** with low limit supply air temperature control, and high limit supply air humidity control.

Fig. 27.9 Air conditioning system diagram. System designed to control the extract air from a room. The air into the space is heated by LTHW, HTHW or a steam heated coil. Cooling and dehumidifying with CHW cooling coil. Humidifying with steam. **Alternative:** Room sensor instead of extract air sensor. **Variants:** with low limit supply air temperature control, and with summer compensation.

Functions

Temperature control

The duct sensor f1 or the room sensor f2 measures the temperature t_R. The controller u1 compares this value with the selected setpoint X_K and adjusts the heating coil valve s1 or the cooling coil valve s2 in sequence in accordance with the difference between the two.

Humidity control

The duct sensor f1 or the room sensor f2 measures the humidity φ_R. The controller u2 compares this value with the selected setpoint X_K and adjusts the humidfying valve s3 or the cooling valve s2 in sequence in accordance with the difference between the two.

Safety devices

When there is danger of frost, the frost protection thermostat f4 must switch off the fan, close the damper s4, open the heating coil valve s1 and, where appropriate, switch on the heating pump.

Variant: with low limit supply air temperature control

The low limit supply air termperature sensor f3 prevents the supply air temperature t_{zu} from dropping below the cut-in point X_E set on the controller u1 (draught elimination).

Variant: with summer compensation.

The outside temperature compensation sensor u3 is used to increase the room temperature t_R in summer. If the outside temperature t_a rises above the cut-in point X_E (22°C), the setpoint X_K is increased continuously by the selected steepness S_S.

Controller output temperature

Controller output humidity

Setpoint curve

Setpoint curve

Engineer uses a psychrometric chart to determine the physical properties of the air to be handled.

Building management

Fig. 27.8 and Fig. 27.9 show possible schemes for partial and full air conditioning. In addition to the controls indicated there will be all the associated ductwork, filters, water, steam and electrical services. An electrical control panel is usually necessary for the wiring of interconnected equipment, instrumentation, and to isolate plant for servicing. Although operations are generally fully automatic, emergency hand control facilities are often specified.

In an industrial situation where many departments exist under the one roof, a central building management system is necessary if the Plant Engineer requires to know what is happening in the installation at any time.

Fig. 27.10 shows a diagrammatic arrangement of a building management system with software designed specifically for the installation. Control is from a standard IBM compatible PC(1), see Fig. 27.11.

The operator can check how any of the peripheral components (4) are working. Control units for major items of equipment such as boilers, fans, etc. are shown as item (3). Individual room controllers (5) control energy consumption as a function of room occupancy, the time of day and season. System controllers (2) coordinate process control tasks such as the overall management of energy.

The Plant Manager has instant access to data using a mouse operation and pull down menus. Individual schematic diagrams can be displayed. Recorded data over a period of time can be displayed or printed out.

The psychrometric chart (Fig. 27.13)

In air conditioning technology, it is necessary to define modynamic processes and the properties of moist air.

This may be achieved by a good knowledge of physics, with theoretical calculations using complicated formulae and tables. The procedure can be time consuming.

By presenting the interrelated factors on a psychrometric chart, an immediate decision can be made regarding the feasibility of controlling an air conditioning system and the means required to carry this out.

For a given air sample nine different parameters are shown on the psychrometric chart.

A position on the chart can be established at the intersection of two ordinates for known conditions and the others obtained.

Since the properties and behaviour of moist air depend on barometric pressure, a psychrometric chart can only be drawn for a specific barometric pressure. Allowances may be made for changes in barometric pressure by using correction factors.

Note that the chart indicates a condition of 21°C Dry bulb temperature and 48% Relative humidity. These are typical values to provide comfort in an office.

Fig. 27.10

Fig. 27.11

Fig. 27.12 Typical displays.
(a) Pull down menus for selecting functions relating to any schematic in the building layout.
(b) Pictorial graph of control operations with shaded bands where limit values have been exceeded.
(c) Data evaluation and display. Electricity consumption in various zones for one month is given in this example.

Fig. 27.13

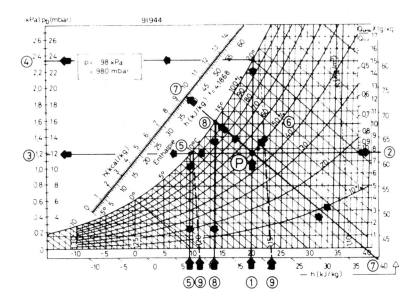

Fig. 27.14

Example: Find the missing values for the following case.

1 Dry bulb temperature	t_{sic}	= 20°C
2 Absolute humidity	$\times$	= g/kg
3 Partial water vapour pressure	p_D	= mbar or kPa
4 Saturation pressure	P_{sat}	= mbar or kPa
5 Saturation temperature (dewpoint)	t_{sat}	= °C
6 Relative humidity	φ	= 50%
7 Enthalpy	h	= kJ/kg
8 Wet bulb temperature	t_{hyg}	= °C
9 Density	ρ	= kg/m3

The point on the chart is defined by the temperature and the relative humidity given above.

Solution:
The point of intersection P between the 20°C isotherm from the dry bulb temperature (1) and the line of 50% constant relative humidity (6) clearly defines the position of the required condition.

The absolute humidity (2) is found by drawing a horizontal line through the point P and extending it until it meets the ordinate on the right.

If this horizontal is extended to the left it will intersect the scale for partial water vapour pressure p_D (3).

To obtain the saturation pressure (4) the isotherm from P must be extended until it intersects the 100% relative humidity line. At this point the air is saturated, i.e. it cannot absorb any further moisture without a dense mist forming. An extension of the horizontal line through this point of intersection to the left intersects the partial pressure scale at point (4). The pressure of the saturated air can now be read.

Where the horizontal line from (P) intersects the saturation curve a similar condition occurs, whereby the air cannot absorb any additional moisture (5). The dewpoint or saturation temperature can now be read on the saturation curve (and also on the dry bulb temperature scale).

By following the isenthalp (line of constant enthalpy) which passes through the condition (P) we can determine the enthalpy at the points of intersection (7) with the enthalpy scale. If an adiabatic line is drawn through the point (P) towards the saturation curve, the two intersect at point (8) to give the wet bulb temperature. This is lower than the starting temperature because the absorption of moisture has caused sensible heat to be converted into latent heat.

The density is determined from the nearest broken lines of constant density (9).

The required values are:

2.	$\times$	= 7,65	g/kg
3.	PD	= 11,8	mbar = 1,18 kPa
4.	P_{sat}	= 23,4	mbar = 2,34 kPa
5.	t_{sat}	= 9,6	°C
7.	h	= 39,8	kJ/kg
8.	t_{hyg}	= 13,8	°C
9.	p	= 1,16	kg/m3

Refrigeration systems and energy-saving applications

In order to appreciate the engineering diagram examples relating to refrigeration practice, we have included an explanation of a typical cycle of operations.

Refrigeration through evaporation

When you pour liquid ether on to the back of your hand, after a few seconds you feel your hand turn ice-cold. The liquid evaporates very quickly, but in order to do so it requires heat—so-called heat of evaporation. This heat is drawn in from all around the ether—including from your hand—and it is this which causes a sensation of cold.

If you could catch this evaporated ether and liquefy it again by compressing and cooling, the heat absorbed during evaporation would be released back into the surroundings.

This is precisely the principle on which the refrigeration cycle works. A special refrigeration agent, which is even more suitable for this purpose than ether, is evaporated close to the medium to be refrigerated. The heat necessary for this process is drawn in from all around, thereby cooling the air or water.

The most widely known refrigerating agent used to be ammonia but this has now been almost entirely superseded by halogen refrigerants, the best known of which are R12, R22 and R502.

The refrigerant cycle

Fig. 27.15 The refrigerant cycle

The refrigerant circulates in a closed system. To produce this circulation, a very powerful pump is required—the compressor. This draws in the expanded refrigerant in vapour form and compresses it. On being compressed, the temperature of the vapour rises. It moves on to the condenser and is cooled by means of cold water. The cooling is so great that the condensation temperature is reached and the condensation heat thus produced is given up to the water in the condenser.

The refrigerant is then pumped on in liquid form into a receiver and from there on to the evaporator. But just before this, it flows through an expansion valve. This valve reduces the pressure on the liquid so much that it evaporates, drawing in the required heat from its surroundings. This is precisely as intended for the air or water to be cooled is led past the group of pipes in the evaporator. After leaving the evaporator, the refrigerant is once again drawn into the compressor.

To summarise: In one half of the cycle, the heat is removed by evaporation (i.e. cooled where cooling is required) and in the other half, heat has been released by condensing. Thus energy (heat) is moved from where it is not wanted to a (different) place where it is tolerated or, in fact, required.

When the theoretical cycle of operations is applied in practice it is necessary to include controls and safety devices. In a domestic system the motor and compressor are manufactured in a sealed housing.

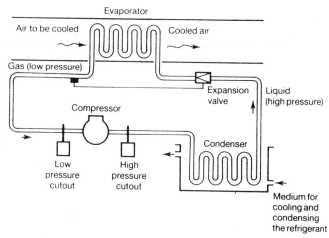

Fig. 27.16 The safety devices in the refrigerant cycle

The heat extracted from the inside of the refrigerator, where the evaporator is positioned, passes to the condenser, generally at the back of the cabinet, where natural convection currents release the heat into the surroundings.

In a large industrial installation it may be economically viable to recover heat from a condenser and use it for another process.

Heat recovery control in refrigeration systems

The principle of the refrigeration cycle is shown in Fig. 27.17. where the refrigerant passes through the following four phases:

1 evaporation (with heat absorption)
2 compression (with energy absorption)
3 condensation (with heat emission)
4 expansion

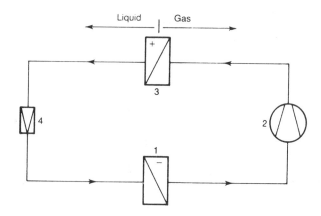

Fig. 27.17

In a refrigeration system, a second condenser can be incorporated in the cycle and the recovered heat used for:

1 heating domestic water,
2 supply air heating,
3 reheating in the case of supply air dehumidification.

Control options

The heat absorbed by the evaporator and that generated by the compressor are emitted in the condenser. To recover this heat an additional condenser can be connected in parallel, and alternative positions of a three way valve give two options: 'Hot gas diversion' or 'Condensate control'.

Fig. 27.18 shows the second condenser fed from a three way diverting valve installed in the hot gas pipe to illustrate hot gas diversion.

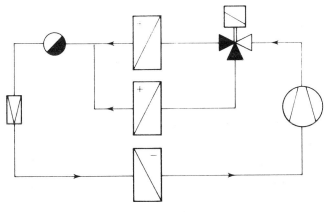

Fig. 27.18

Option 1 When the controller transmits a demand for heating (e.g. via the supply air temperature in the air conditioning system), the modulating diverting valve opens the flow to the heat recovery condenser and closes the supply to the other condenser. If not all the heat recovered is required, the remainder is dissipated via the second condenser.

Fig. 27.19 shows the connections for condensate control with the three way value positioned in the liquid pipe.

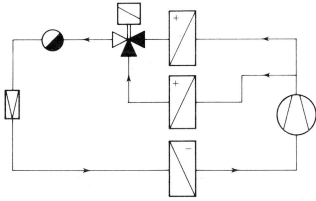

Fig. 27.19

Option 2 The entire refrigerant flow is normally directed through the valve (see Fig. 27.19). The three-way liquid valve opens the condensate pipe of the heat recovery condenser and closes the pipe of the second condenser when the heat recovery condenser has to emit effective heat. This condenser then releases all condensate thus making its heat exchange surface free for condensation of fresh hot gas. At the same time, the second condenser is filled with liquid thus making its exchange surface inactive. When the demand for effective heat decreases, the process is reversed.

The following examples show how the two options have been applied in a swimming pool/ice rink and a covered shopping complex.

The system shown in Fig. 27.20 relates to an air conditioning system in an indoor swimming pool combined with a low temperature refrigeration system for an ice rink. It is designed on the principle of the schematic in Fig. 27.18 (hot gas diversion) in order to provide a fast response time of the supply air control.

Key

1 Control valve
2 Expansion valve
3 Evaporator
4 Compressor
5 Heat recovery condenser
6 Outside condenser

Fig. 27.20   Indoor swimming pool/ice rink

Indoor swimming pools normally require large quantities of heat and have high relative humidity. If cold outside air is mixed with the humid return air, condensation results. To avoid this, the outside air is preheated by a heat recovery condenser (5).

Systems installed in enclosed shopping complexes have the following characteristics:
—large heat demand for air conditioning of shopping street
—large heat output from refrigerated cabinets.

The two condensers in the system illustrated are

Key

1 Control valve   4 Compressor
2 Expansion valve   5 Heat recovery condenser
3 Evaporator   6 Outside condenser

Fig. 27.21 Air conditioning and refrigeration system in an enclosed shopping complex

connected in series. The heat recovery condenser (5) can be switched off completely in summer. In winter, spring and autumn it is used to pre-heat the outside in accordance with demand.

Fig. 27.22 shows a simplified schematic of the refrigeration cycle.

The authors wish to express their thanks for the assistance given and permission to include examples of applications engineered by Staefa Control System Ltd, Hawthorne Road, Staines, Middlesex.

Key

1 Control valve   4 Compressor
2 Expansion valve   5 Heat recovery condenser
3 Evaporator   6 Outside condenser

Fig. 27.22 Refrigeration cycle of the shopping complex (simplified)

Pneumatic systems

Pneumatic systems require a supply of clean compressed air to motivate cylinders, tools, valve gear, instruments, delicate air controls and other equipment. Most factory and plant installations operate between 5.5 and 7 bar.

A typical compressor installation is shown in Fig. 27.23.

Compressors are sized according to the amount of free air delivered. Air flow is measured in cubic decimetres per second (dm^3/s) at standard atmospheric conditions of 1013 mbar and 20°C as specified in ISO 554. The compressed air is stored in an air receiver and, for a system operating at pressures in the region of 7 bar gauge, the size of the receiver in litres should be approximately equal to 30 times the rated free air delivery of the compressor in dm^3/s. Thus, a compressor rated at 50 dm^3/s free air delivery requires a receiver of approximately 1500 litres capacity.

Compressed air in any normal supply mains contains contaminants which need to be either completely or partially removed depending on the ultimate use of the compressed air. Naturally, the cleaner the air has to be the greater the expense. The contaminants are:

a) water in liquid and vapour form;
b) oil which can exist in three forms, oil/water emulsions, minute droplets suspended in the air as an aerosol and oil vapours;
c) atmospheric dirt particles and solid particles formed by the heat of compression.

Having considered the types of contaminant present in an air system, one can decide upon the degree of cleanliness needed for any particular process and the means required to obtain this. These include After-coolers, Receivers, Air line filters, Air dryers, Coalescing filters, Vapour adsorbers and Ultra high efficiency dirt filters. Each application must be considered on its merits. Fig. 27.24 shows a typical air line installation for a factory. Further cooling may occur in the distribution mains themselves; these should be installed with a pitch in the direction of the air flow, so that both gravity and air flow will carry water to drain legs at appropriate intervals. These legs should be fitted with automatic drain valves. Note that take-off points are connected to the top of the distribution mains to prevent water entering the take-off lines.

The quality of air required for plant use will now dictate which accessories are to be fitted at each take off point. These range from a selection of filters and pressure regulators; and if lubrication is required in air actuated components, then lubricant can be metered and atomized in the air line in the form of a fine fog to coat all operating parts with a thin protective film. The equipment is lubricated automatically through its operating cycle.

Regular maintenance will ensure trouble-free production facilities.

Industrial processes include: air agitation, air bearings, air conveying of foodstuffs and powders, air motors i.e. rotary, reciprocating and linear air cylinders, blow guns, cleaning and cooling nozzles, breathing masks and protective clothing, fluidics, food and drink processing, general machinery, instrumentation, pneumatic circuits and valves, and spray guns.

Fig. 27.23

1. Air Intake Filter
2. Air Compressor
3. Air Compressor Water to Air Heat Exchanger
4. Air Receiver
5. Safety Valve
6. Isolating Valve
7. Main Line Air Filter
8. Automatic Air Receiver Drain
9. Drip-Leg Drain
10. Dryer
11. 'Puraire' After-Filter
12. Filter-Regulator
13. Automatic Drain Filter
14. Lubricator
15. 'Ultraire' Filter*
16. Precision Controller*
17. Primary Filter*
18. Precision Regulator*
19. 'Ultraire' Filter

A. Pitch with Flow
B. To Machine Shop
C. Wide Pattern Return Bends
D. To Gauging Equipment
E. Dry Air to Process Control
F. Olympian 'Plug-In' Vitalizer Unit
G. Olympian 'Plug-In' Lubro-Control Unit
H. To Future Extensions

*Precision Air Sets

Fig. 27.24

Fig. 27.25. Circuit symbols can be drawn, stored in the computer database, and used repeatedly on other diagrams. Neat layout work manually often involves changes and a lot of repositioning. This is time consuming, but the computer handles alterations with speed and accuracy.

Pneumatic circuit design

The first requirement in circuit design is a thorough understanding of the symbols to be used. The most important and frequently used symbols are the five port and three port valves. They are also the most frequently misunderstood symbols, and therefore we start by showing the build-up of a typical five port valve. In Fig. 27.26. a double-acting cylinder is shown connected to a five port valve. The square envelope represents the valve body and there are three ports on the bottom

Fig. 27.25 Circuit symbols can drawn, stored in computer database, and used repeatedly on other diagrams. Neat layout work manually often involves changes and a lot of repositioning. This is time consuming, but the computer handles alterations with speed and accuracy

Table 27.2   Selected symbols for fluid power systems (from BS 2917)

NOTE 1. The symbols for hydraulic and pneumatic equipment and accessories are *functional* and consist of one or more basic *symbols* and in general of one or more *functional symbols*. The symbols are neither to scale nor in general orientated in any particular direction.

NOTE 2. In circuit diagrams, hydraulic and pneumatic units are normally shown in the unoperated position

NOTE 3. The symbols show connections, flow paths and the functions of the components, but do not include constructional details. The physical location of control elements on actual components is not illustrated.

Description	Symbol		Description	Symbol	Description	Symbol
General symbols			*Air-oil actuator (transforms pneumatic pressure into a substantially equal hydraulic pressure or vice versa)*		spring loaded (opens if the inlet pressure is greater than the outlet pressure and the spring pressure)	
Basic symbols						
Restriction:			Directional control valves		pilot controlled (opens if the inlet pressure is higher than the outlet pressure but by pilot control it is possible to prevent:	
affected by viscosity			*Flow paths:*			
unaffected by viscosity			one flow path			
Functional symbols			two closed ports		closing of the valve	
hydraulic flow	▼		two flow paths			
pneumatic flow or exhaust to atmosphere	▽		two flow paths and one closed port		opening of the valve)	
Energy conversion			two flow paths with cross connection			
Pumps and compressors					with resistriction (allows free flow in one direction but restricted flow in the other)	
Fixed capacity hydraulic pump: with one direction of flow			one flow path in a by-pass position, two closed ports			
with two directions of flow						
Motors			*Directional control valve 2/2:*			
Fixed capacity hydraulic motor: with one direction of flow			with manual control		*Shuttle valve (the inlet port connected to the higher pressure is automatically connected to the outlet port while the other inlet port is closed)*	
			controlled by pressure against a return spring			
Oscillating motor: hydraulic			*Direction control valve 5/2:*		Pressure control valves	
Cylinders			controlled by pressure in both directions		*Pressure control valve:*	
Single acting:	Detailed	Simplified			one throttling orifice normally closed	
returned by an unspecified force			NOTE. In the above designations the first figure indicates the number of ports (excluding pilot ports) and the second figure the number of distinct positions.		one throttling orifice normally open	
Double acting: with single piston rod			Non-return valves, shuttle valve, rapid exhaust valve		two throttling orifices, normally closed	
Cylinder with cushion: single fixed			*Non-return valve:* free (opens if the inlet pressure is higher than the outlet pressure)		*Sequence valve (when the inlet pressure overcomes the spring, the valve opens, permitting flow from the outlet port)*	

Table 27.2 *(continued)*

Description	Symbol	Description	Symbol	Description	Symbol
Flow control valves		Rotary connection		Over-centre device (prevents stopping in a dead centre position)	
Throttle valve: simplified symbol		one way			
Example: braking valve		three way		Pivoting devices: simple	
Flow control valve (variations in inlet pressure do not affect the rate of flow): with fixed output	* Simplified	Reservoirs		with traversing lever	
		Reservoir open to atmosphere:		with fixed fulcrum	
Flow dividing valve (divided into a fixed ratio substantially independent of pressure variations)		with inlet pipe above fluid level		Control methods	
Energy transmission and conditioning		with inlet pipe below fluid level		*Muscular control:* general symbol	
Sources of energy		with a header line		by push-button	
Pressure source		*Pressurized reservoir*		by lever	
		Accumulators		by pedal	
Electric motor		*The fluid is maintained under pressure by a spring, weight or compressed gas*		*Mechanical control:* by plunger or tracer	
				by spring	
Heat engine		Filters, water traps, lubricators and miscellaneous apparatus		by roller	
		Filter or strainer		by roller, operating in one direction only	
Flow line and connections		Heat exchangers		*Electrical control:* by solenoid (one winding)	
Flow line: working line, return line and feed line		*Temperature controller (arrows indicate that heat may be either introduced or dissipated)*		by electric motor	
pilot control line		*Cooler (arrows indicate the extraction of heat)*		*Control by application or release of pressure*	
drain or bleed line		with representation of the flow lines of the coolant		*Direct acting control:* by application of pressure	
flexible pipe		*Heater (arrows indicate the introduction of heat)*		by release of pressure	
Pipeline junction		Control mechanisms		*Combined control:*	
Crossed piplines (not connected)		Mechanical components		by solenoid and pilot directional valve (pilot directional valve is actuated by the solenoid)	
Air bleed		*Rotating shaft:* in one direction		Measuring instruments	
Power take-off plugged		in either direction		*Pressure measurement:* pressure gauge	
with take-off line		*Detent (device for maintaining a given position)*		Other apparatus	
connected, with mechanically opened non-return valves		*Locking device (* symbol for unlocking control)*		*Pressure electric switch*	
uncoupled, with open end					
uncoupled, closed by free non-return valve					

Piston rod in the 'minus' position

Fig. 27.26

Piston rod in the 'plus' position

Fig. 27.27

edge and two on the top edge. A compressed air supply is connected to the centre port 1. Air exhausts to atmosphere at ports 3 and 5. Air outlets to power the cylinder at ports 2 and 4. The lines within the envelope show the passages within the valve for the current valve state. The air supply 1 is connected to outlet 4 and outlet 2 is connected to exhaust 3. Exhaust 5 is sealed. This means that the cylinder has air pressure pushing the piston to the instroked or 'minus' position. The other side of the piston is connected to exhaust.

To make the cylinder move to the outstroked or 'plus' position the valve has to be operated to change to its new state. This is shown in Fig. 27.27. Note that the envelope and port connections are exactly the same and it is only the connection paths inside the valve that have changed.

The full symbol for a 5/2 valve (five ports, two positions) are these two diagrams drawn alongside each other. Only one half will have the ports connected.

Which half, will depend on whether the cylinder is to be drawn in the instroked or outstroked state. The method by which the valve is operated, i.e. push button, lever, foot pedal etc. is shown against the diagram of the state that it produces.

Fig. 27.28 shows a 5/2 push button operated valve with spring return. It is operating a double acting cylinder. In addition a pair of one way flow regulators are included to control the speed of piston rod movement. The symbol for this type of flow regulator consists of a restrictor, an 'arrow' which indicates it is adjustable and a non return valve in parallel, to cause

restriction in one direction only. The conventional way to control the speed of a cylinder is to restrict the exhausting air. This allows full power to be developed on the driving side of the piston which can then work against the back pressure and any load presented to the piston rod.

Study Fig. 27.29 and imagine that when the push button is pressed the complete symbol moves sideways to the left, but leaves the pipe connections and port numbers behind so that they line up with the other half of the diagram. In this position the cylinder piston rod will move out to the 'plus' position. Imagine the spring pushing the symbol back again when the button is released. The numbers at the valve ends signify which output will be pressurized when the valve is operated at that end. If the button is pushed at end 12 then port 1 will be connected to port 2.

If the button is released, the spring at end 14 becomes dominant and port 1 will be connected to port 4.

A three port valve symbol works in a similar way. Two diagrams of the valve are drawn side by side. Fig. 27.29 shows the full symbol for a 3/2 valve controlling a single acting cylinder. Port 1 is the normal inlet, port 2 the outlet and port 3 the exhaust. The valve end numbers 12 and 10 indicate that port 1 will be con-

Fig. 27.28

Fig. 27.29

nected either to 2 or to 0 (nothing). Since there is only one pipe supplying a single-acting cylinder, speed control of the 'plus' motion has to be obtained by restricting the air into the cylinder. Speed of the 'minus' motion is effected conventionally by restricting the exhausting air. To provide independent adjustment two one-way flow regulators are used and these are connected in the line back-to-back.

Logic Functions

Designers of pneumatic circuits are not usually consciously thinking in pure logic terms, but more likely designing intuitively from experience and knowledge of the result that is to be achieved. Any circuit can be analysed however, to show that it is made up of a combination of logic functions. The four most commonly used are illustrated in Figs. 27.30 to 27.33.

The AND function. The solenoid valve A (AND) the plunger operated valve B must both be operated before an output is given at port 2 of valve B. Fig. 27.30.

The OR function. For this a shuttle valve is required so that either of two push-button valves A (OR) B can provide a signal that is directed to the same destination. The shuttle valve contains a sealing element that is blown by the incoming signal to block off the path back through the other valves' exhaust port. Fig. 27.31.

The NOT function. This is simply a normally open valve. When it is operated by a pilot signal on port 12 it will NOT give an output. The outlet will be given when the valve re-sets to its normal state by removing the signal. Fig 27.32.

The MEMORY function. When a double pressure operated three port valve is given a signal at port 12, an output is obtained at port 2. If the signal is now removed the output will remain, it has REMEMBERED its ON state even when the signal that caused it has gone.

If a signal is given to port 10 the valve will re-set and the output exhausted. If the signal is removed the new OFF state is REMEMBERED. Fig. 27.33.

The TIME DELAY. By using a flow regulator and a 3/2 pilot operated pressure switch, a signal can be slowed down to provide a time delay. Fig. 27.34 shows that when a signal is fed through the flow regulator, it will slowly build up pressure in an air reservoir (R) and on the signal port 12 of the pressure switch. This will continue until the pressure is high enough to operate the pressure switch. Then, a strong unrestricted signal will be sent to operate a control valve or other device. The delay can be adjusted by changing the setting on the flow regulator. A reservoir, of approximately 100 cc in volume, would allow a delay range of between 2 and 30 seconds. Without the reservoir, the range will be reduced to approximately 3 seconds maximum. Note that the pressure switch is like a pilot operated 3/2 valve, but uses air pressure as a return spring. The pilot signal on port 12 overcomes this, as it is working on a larger area piston.

Fig. 27.30

Fig. 27.31

Fig. 27.32

Fig. 27.33

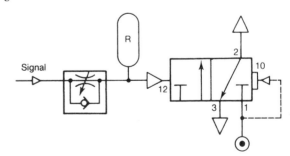

Fig. 27.34

A semi-automatic circuit is shown in Fig 27.35 When the push button is operated and released, the 3/2 valve will send a signal to operate the 5/2 double pilot valve. This will cause the cylinder to move to the 'plus' position. A cam on the piston rod will operate the roller

Fig. 27.35

plunger valve and this will give a signal to re-set the 5/2 valve. The piston rod will then automatically move to the 'minus' position and wait until a further operation of the push button is given.

Sequential circuits

In an automatic system where two or more movements are to occur in a specific order, a sequence is formed. A typical example is a special purpose automatic machine. This may be carrying out a manufacturing, or packaging operation where air cylinders are used to power the movements in a continuously repeating sequence.

Each movement in a sequence can be produced by a pneumatic cylinder. This will either be single acting, or double acting and the choice depends on whether there is any return resistance or load requiring a powered return. Single acting cylinders are controlled by a 3/2 double pilot operated valve and double acting cylinders are controlled by a 5/2 double pilot operated valve.

For each cylinder used, a circuitry building block can be established. See Fig. 27.36. This illustrates a double acting cylinder building block for the cylinder labelled

'A'. Two command signals are required, one to move it 'plus' (a+), the other to move it 'minus' (a−). To prove that the movements have been completed, two feedback signals are required. These are provided by the two roller operated 3/2 valves. One proving the 'plus' movement (a1), the other proving the 'minus' movement (a0).

Consider a two cylinder system where the cylinders are labelled A and B. The sequence required after selecting the RUN control is $A+B+A-B-$, it will then repeat continuously until the operator selects the END control. The circuit is constructed from two building blocks. See Fig 27.37. Note that flow regulators are included in the power lines to each end of the cylinders.

These provide adjustable speed control for each movement. To RUN and END the repeating cycle a 3/2 manually operated valve is included.

The two building blocks form a complete circuit by having their command and feedback lines connected together. The method of interconnection is achieved by application of this simple rule:

'The proof of position signal resulting from the completion of each movement is connected to initiate the next movement.'

The circuit can be traced as follows:

Start with the output given from the RUN/END valve when it is switched to RUN.

The a+command is given.

Cylinder A moves+.

The a1 proof of position signal results.

This becomes the b+command.

Cylinder B moves+.

The b1 proof of position signal results.

This becomes the a − command.

Cylinder A moves − .

The a0 proof of position signal results.

This becomes the b − command.

Fig. 27.36

Fig. 27.37

Cylinder B moves −.

The b0 proof of position signal results.

This becomes the supply to the RUN/END valve.

If the RUN/END valve is still switched to RUN a repeat cycle will be started.

This simple daisy chain method of interconnection will work for any number of cylinders, provided the sequence allows their return movements to occur in the same order as their first movements. For this to be true, the first movement of a cylinder need not be plus nor is it necessary for the first half of the sequence to be in alphabetical order, eg the sequence B + A − D + C- − B − A + D − C+ conforms to these rules and can be solved with this simple daisy chain method.

If the cylinders do not return in exactly the same order as their first movements complications will arise. Take for example, the sequence A + B + B − A − and repeat. If we try to interconnect the equipment for this

sequence in the same way as before, there will be two states where the 5/2 valves will have both a 'plus' and 'minus' command existing at the same time, therefore preventing operation. This condition is commonly known as opposed signals and can be cured in a variety of ways. For the most reliable and economical method we suggest the use of the Cascade system. See Fig 27.38.

The cascade technique is to switch on and off the supply air to the critical trip valves in groups. The need for this will occur when a trip valve's mechanism is still held down, but the output signal has been used and requires removing. By switching off the group air that is supplying the valve, the output is also removed and achieves the desired result. After the valve's mechanism is naturally released in the sequence, the group supply is switched on again in time for its next operation. To determine the number of cascade groups for

Fig. 27.38

any sequence, the sequence must be split into groups starting at the beginning, so that no letter is contained more than once in any group. The group numbers are given roman numerals to avoid confusion with other numbering systems that may exist on larger systems. The placing of the RUN/END valve should be in the line that selects group I. This determines that the first task of group I is to signal the first movement of the sequence. In addition, when the circuit is at rest, inadvertent operation of an uncovered trip valve will not risk an unwanted operation of a cylinder.

By studying Fig. 27.38 it can be seen that the sequence splits into two groups. These groups are supplied from a single, double pressure operated 5/2 valve, so that only one group can exist at any time. This is known as the cascade valve.

It can also be seen that neither of the 5/2 valves controlling the cylinders can have the + and − command lines as opposed signals, since their source is from different groups.

The circuit can be traced as follows:

To start, set RUN/END valve to RUN. This generates a command to select group I.

Group I gives a command a+.

Cylinder A moves+.

Valve a1 is operated and generates a command b+.

Cylinder B moves+.

Valve b1 is operated and generates a command to select group II.

Group II gives a command b− (because group I has been switched off there is no opposing signal from a1).

Cylinder B moves −.

Valve b0 is operated and generates a command a− (no opposed signal).

Cylinder A moves −.

Valve a0 is operated and generates a command to start the sequence again.

If at any time the RUN/END valve is switched to END, the current cycle will be completed, but the final signal will be blocked and no further operation will occur.

The rules for interconnection are as follows:

1 The first function in each group is signalled directly by that group supply.
2 The last trip valve to become operated in each group will be supplied with main air and cause the next group to be selected.
3 The remaining trip valves that become operated in each group are supplied with air from their respective groups and will initiate the next function.

Pneumatics and electronics

Systems of low complexity and those in use in hazardous areas, not compatible with electronics, will probably be designed as pure pneumatic systems.

A purely pneumatic system can be viewed as three main sections:

1 Generation and preparation of the compressed air source.
2 Power actuation of pneumatic cylinders through directional control valves.
3 Pneumatic signal processing or logic control.

Electronics can influence all of these sections, for example:

a) By electronic management control of compressors and controlled pressure regulation.
b) In section 2 there are solenoid valves that provide proportional flow and pressure, together with air cylinders having electronic proportional feedback.
c) In section 3, for many systems pneumatic logic has been replaced completely by electronic sequence or logic control.

Programmable sequence controllers (sequencers) and programmable logic controllers (PLCs) are commonly used devices and offer a wide range of features such as timing, counting, looping and logic functions. If a proposed scheme involves a sequence of events more complicated than that shown in Fig 27.38, then electronic possibilities should be explored. In addition to sequence operations there may be the additional complications from long counting operations, or a number of time delays, requiring a high degree of repeatable accuracy. Here the electronic controller will usually be the better choice. Inputs to the controller indicate the completion of the cylinder movement.

These are most conveniently achieved by using a magnetic cylinder fitted with reed switches. The reed switch consists of two spring like metal reeds within a sealed enclosure. When a magnet around the piston is within range, the reeds are magnetized, each having a N and S pole. As the free ends will be of the opposite polarity they snap together. For environments where strong magnetic fields exist mechanical limit switches may be used.

The scope of such a system will be appreciated from Fig. 27.39. Programming methods vary with the type of controller and for someone with no experience it is generally easier than they think. Sequencers are designed to be easy to program and are a good choice for machines where the actions are performed in a one-after-the-other interlock. Sequencers are able to jump from one part of the sequence to another, run sections of a sequence in a repeating loop, time, count and perform logic functions such as AND, OR, NOT, etc. It may also be possible to hold several sequences in a memory and select the desired one for a particular task. Sequencers will have a built in range of control buttons to provide facilities such as, run/end cycle, emergency stop, single cycle, auto cycle and manual over-ride.

It takes a little longer to program a PLC. This is produced by keying in a list of logic statements first determined by drawing a ladder diagram. A ladder diagram for a PLC is a logic circuit of the program as it relates to a machines function and sequence. The ladder diagram illustrated in Fig 27.40 is derived from, and similar to the ladder electrical circuits used to

Fig. 27.39

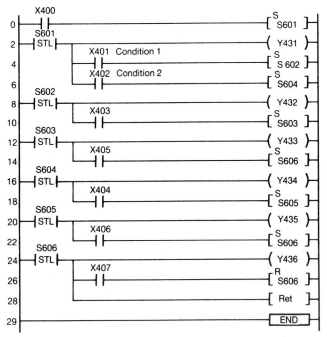

Fig. 27.40

design electro mechanical relay systems.

Pneumatic and electronic systems play an important part in production engineering and typical applications are the control of the main axes of variable pick and place arms and robotics.

The authors wish to express their thanks for the assistance given and permission to include examples of applications of pneumatic controls manufactured by Norgren Martonair Limited, Campden Road, Shipston-on-Stour, Warwickshire.

BS 1553 specifies graphical symbols for use in general engineering. Within the European Community, many additional symbols are in common use and a selection of these are included here for reference purposes.

Table 27.3

	Equipment Labelling
	positioning or flow direction
	steps
	on (IEC 5007)
	off (IEC 5008)
	stand-by (stand-by position IEC 5009)
	day (normal mode), sun, brightness (IEC 5056)
	cooling, frost, cold (below 0°C) (ISO 0027)
	heating flame (ISO 0364)
	wind, wind influence (international)
	manual actuation, manual control (ISO 0096)
	safety insulation, electrical protection Class II (IEC 5172)
	extra-low voltage (up to 50 volts), protection Class III (IEC 5180)
	dangerous electrical voltage (voltage indication) (IEC 5036)
	flow arrow, indicating passage, entry and exit of important substances (ISO 0251)
	indicator arrow (ISO 0952)
	temperature increase, increase temperature (ISO 0035)
	temperature reduction, reduce temperature (ISO 0036)

Table 27.3 *(continued)*

Systems symbols		Systems symbols	
	heating energy, energy demand (ISO 0523)		cam control (IEC 117-3)
	heat exchanger, general with substance flows crossing (ISO 0110)		electromagnetic actuator (DIN 30600)
	heat exchanger, general without substance flows crossing (ISO 0111)		example: magnetic valve showing flow
	tank, general (pressureless) (ISO 0101)		liquid pump, circulating pump, general (ISO 0134)
	tank with convex bottom, general (for high pressure) (ISO 0103)		fan (general) (~ISO 1219)
	isolating valve (general) two way valve (ISO 0234)		compressor (general) (ISO 0137)
	three way valve (ISO 1219)		compressor, 4 step
	four way valve (ISO 1219)		air damper (~ISO/TC59)
	flow symbols: (~DIN 19227)		air filter (general) (~DIN 1946E)
	– variable		
	– constant		heating coil (~DIN 1946E)
	mixing		
	diverting		cooling coil (~DIN 1946)
	shower, nozzle (ISO 0145)		device or function unit, general (IEC 117-13)
	steam trap (ISO 0157)		modulating controller (general) (IEC 5183)
	filter (ISO 0114)		keys, keyboard (IEC 5211)
	manual actuator (IEC 117-3)		sensor with on-off function (e.g. thermostat, hygrostat pressure switch etc.) (IEC 298)
	self operated actuator (or actuator in general) (IEC 117-3)		sensor with on-off function (e.g. thermostat, hygrostat pressure switch etc.) with immersion, duct or capillary pocket (IEC 298)
	electromotoric actuator (ISO 0147)		
	electrothermic actuator (ISO 1219)		immersion thermostat for temperature (IEC 298/1)
	hydraulic or pneumatic actuator (IEC 117-3)		
	diaphragm actuator (DIN 30600)		

Other references examples:

x	absolute humidity	h	enthalpy
p	pressure	aq	air quality (SCS)
Δp	differential pressure		occupancy
$\dot{V}$	flow, volume flow rate		etc.
v	velocity		

Table 27.3 (continued)

IEC 117-... (DIN 40711)	Symbols for electrical schematics		Symbols for electrical schematics
———	DC-current, also DC-voltage (general)		combined wires, general, any sequence on each side (wires should be coded)
— — —	alternative (use this symbol only where there is a risk of confusion on diagrams) (IEC 5031)		combined wires, general, as above but single line representation
~	AC-current, also frequency in general AC-voltage (frequency indicated where necessary – on the right of the symbol, e.g. ~50 Hz) (IEC 5032)		general symbol denoting a cable
≂	suitability for use on either DC or AC supply (IEC 5033)		example: 2 core cable
+	positive polarity (IEC 5005)		example: 2 core cable 'screened' (general)
—	negative polarity (IEC 5006)		coaxial line, screened
	definitions of electrical conductors in accordance with IEC standard 364-3B (CENELEC) (DIN 40719)		crossing of conductor symbols no electrical connection
	– L Phase (formerly PH) – N Neutral (formerly N) – L₁ Phase 1 (formerly R) – L₂ Phase 2 (formerly S) – L₃ Phase 3 (formerly T) – PE Earth		junction of conductors
m ~ f U	AC-current with m phases, frequency f and voltage U		general contact, in particular one that is not readily separable e.g. soldered joint
			readily separable contact e.g. terminal on controller base
3N ~ 50 Hz 380V	Example: three-phase AC-current with neutral wire, 50 Hz, 380 V (220 V between phase wire and neutral wire)		terminals: (IEC 117-3, DIN 40711)
			device terminals
———	one wire or a group of wires		control panel terminals: (IEC 240)
∿	flexible wires	U > 50 V	– on connection diagram
—⫴—	line showing the number of wires e.g. 3 wires	U < 50 V	
—ⁿ⁄—	numbers of wires = n		
—⁸⁄—	example: 8 wires	U > 50 V O U < 50 V ⦸	– on circuit diagram
—‖—	line showing the number of circuits e.g. 2 circuits		plug or plug pin (IEC, ISO 0345)
	Combining wires for the sake of simplicity in wiring diagrams		socket outlet (IEC, ISO 0345)
			fuse general (IEC 5016) (aspect ratio on drawing 1:3)

Table 27.3   *(continued)*

Symbols for electrical schematics		Symbols for electrical schematics	
	fuse showing supply side (IEC 117-3)		– anti-clockwise
	voltage fuse general over voltage discharge device surge arrestor (IEC 117-3)		– both directions
	isolating point with plug-in connection (IEC 117-3, DIN 40713)		thermostat, hygrostat etc. e.g. p → pressure switch
	earth, general (IEC 5017)		Manually operated control, general (IEC 117-3)
			– this symbol is used when space is limited
	safety conductor, safety earth (IEC 5019) protection class I		manual operation by pushing
	chassis, general (IEC 5020)		manual operation by pulling
	GND (ground, common chassis)		manual operation by turning
	resistor, general (IEC 117-1) (aspect ratio on drawing 1 : 2.5)		manual operation by toggle or lever
	inductor, inductive reactance (1 : ⩾ 2) (IEC 117-1)		actuator general, e.g. for relay, contactor (IEC)
	capacitor, capacitive reactance (ratio approx. 1 : 4) (IEC 117-1)		electromechanical actuator, e.g. showing active winding (IEC)
	polarised (electrolytic) capacitor (IEC 117-7)		electromechanical actuator with two windings active in the same direction (IEC)
	motor, general (ISO 0147)		signal lamp general 'operation' (IEC 5115, DIN 40708)
	transformer with two separate windings (IEC 5156)		signal lamp, flashing for fault (DIN 40708)
	as above (alternative representation) (IEC 500)		signal lamp 'fault', emergency lamp (≈IEC, ≈DIN 40717)
	battery of cells or accumulators (the long line represents the positive pole) (IEC 5001)	(IEC 117-3)	
	mechanical coupling: (IEC 117-3, DIN 40703)		buzzer
	– general symbol		bell
	– symbol used when space is limited		horn (IEC 5014)
	linear motion: (IEC 117-3)		siren
	– to the right		transducer, signal transducer, transmitter, general symbol (IEC 117-3)
	– to the left		
	– both directions (IEC 5023) (limited in both directions)		
	rotational motion: (IEC 117-3)		
	– clockwise		

Table 27.3 *(continued)*

Symbols for electrical schematics		Symbols for electrical schematics	

 rectifier, rectifying device general (IEC 5186)

 anplifier general symbol (IEC 5084)

 oscillograph, general symbol (IEC 177-4)

 recording measuring device, recorder (IEC 5443)

 recording measuring device, printer (IEC 5444)

 remote operation, general (~ISO 0093)

 adjuster (IEC 5285)

 communication (electronic) – ISO (from IBM system, communication LINK)

 clock, general (IEC 117-4)

 synchronous clock (IEC 117-4)

 time clock (IEC 117-4)

 dividing line (e.g. between two zones or to separate a space) (IEC 117-1)

 example: control panel (IEC 117-1) (adapt line intensity to suit optical effect)

(IEC)

 semi-conductor rectifier diode (IEC = 5186)

 zener diode ($\hat{=}$IEC)

 PNP-transistor   E– –emitter
($\hat{=}$IEC)   C – collector
                            B – base

 NPN-transistor
the collector is connected to the housing ($\hat{=}$IEC)

 optocoupler (SCS) combined symbol based on IEC

Variability (IEC 117-6, DIN 40712)

 inherent linear variability under influence of a physical variable

 inherent non-linear variability under influence of a physical variable

 continuous variability by mechanical adjustment, general

 adjustable in steps

 non-inherent non-linear variability

 continuous variability by mechanical adjustment, linear

 continuous variability by mechanical adjustment, non-linear

 pre-set mechanical adjustment, general symbol

Example:
temperature dependent resistor with negative temperature coefficient (thermistor)

Wire colour abbreviations:

bl	blue	ws	white
dbl	dark blue	sw	black
hbl	light blue	og	orange
rt	red	vl	violet
gb	yellow	gb/gn	yellow/green
gn	green	bn	brown
gr	grey		

International colour code

Colour reference for resistance value and its tolerance

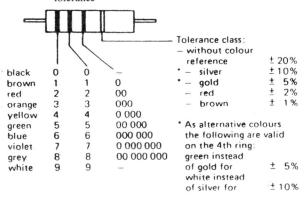

				Tolerance class:
				– without colour reference ± 20%
black	0	0	–	• – silver ±10%
brown	1	1	0	• – gold ± 5%
red	2	2	00	– red ± 2%
orange	3	3	000	– brown ± 1%
yellow	4	4	0 000	
green	5	5	00 000	* As alternative colours
blue	6	6	000 000	the following are valid
violet	7	7	0 000 000	on the 4th ring:
grey	8	8	00 000 000	green instead
white	9	9	–	of gold for ± 5%
				white instead
				of silver for ±10%

e.g. brown – green – red – gold
  1   5   00 = 1500 Ω ± 5%

Chapter 28
Bearings and applied technology

When surfaces rotate or slide, the rotational or sliding motion results in friction and heat. Energy is used, the surfaces wear, and this reduces component life and product efficiency. Friction may be reduced by lubrication which keeps the surfaces apart. At the same time, lubricants dissipate heat and maintain clean contact surfaces. Materials are carefully selected with appropriate mechanical and physical properties for bearings and their housings, to minimize the effects of friction, and particular care is taken with the accuracy of machining, surface finish and maintenance of all component parts associated with bearings.

In a plain bearing the relative motion is by sliding in contrast with the rolling motion of ball and roller bearings.

Plain bearings

Plain bearings may be classified as follows:

The bearing metal should have a low coefficient of sliding friction, be able to conduct heat generated away from the bearing surfaces, resist wear in use and be tough enough to withstand shock loading in service. In the event of breakdown due to lack of lubrication, it may be desirable when overheating occurs for the bearing material to run, preventing seizure and possible severe damage to associated mechanical parts.

Direct lined housings

These housings are lined directly with bearing materials and the choice of material is limited by the practicality of keying or bonding the bearing material to the housing surface.

The dimensions of the housings, casting temperatures and bonding characteristics of the bearing materials will determine whether metallurgical bonding is possible without unacceptable distortion of the housing. Generally this technique is limited to ferrous housings with low melting point whitemetal bearing surfaces. Light alloy and zinc base housings are difficult to line directly with whitemetal.

Insert liners

These are bearing elements which consist of a liner inserted into a previously machined housing and they can be divided into separate classes:

a) Solid insert liners.
b) Lined inserts.
c) Wrapped bushes.

Solid insert liners   Manufactured wholly from suitable bearing materials such as aluminium alloy, copper alloy or whitemetal, these liners consist of machined bushes, half bearings and thrust washers.

The housings are machined to relatively close tolerances. An insert may be finished machined after assembly or a prefinished standard precision liner added as a final operation and this has the added advantage of spares replacement.

Typical applications of insert liners are to be found in diesel engine small bores, crank shaft main bearings, bushes for gearboxes, steering gear and vehicle suspensions.

Lined inserts   These consist of a backing material such as cast iron, steel or a copper alloy which has been lined with a suitable bearing surface of aluminium or copper alloy, or of whitemetal. This type can also be supplied as a solid insert, a split bush, half bearing or thrust washer.

Insert bearing half liners are manufactured as:

a) Rigid or thick walled bearings.
b) Medium walled bearings.
c) Thin walled bearings.

Thick walled bearings   These are backing shells of cast iron, steel pressings and copper base alloys generally lined with whitemetal and copper alloys are used to produce bearings which are manufactured as pairs and used in turbines, large diesel engines and heavy plant machinery. Usually more economic than direct lined housings, these bearings may be provided with a finishing allowance for the bore and length which is adjusted during assembly.

Medium walled insert liners Normally a steel backing is used with a wide range of lining materials. Bearings are prefinished in bore and length and manufactured as interchangeable halves.

Thin walled insert liners These are high precision components with steel backing and whitemetal or copper and aluminium base alloy surfaces, and are suitable for universal application in large production products such as high speed diesel engines and compressors.

Wrapped bushes These are pressed from flat strip of rolled bronze, or steel lined with whitemetal, lead bronze, copper-lead, or aluminium alloys. They are supplied as a standard range of prefinished bushes or with a bore allowance for finishing in situ by fine boring, reaming, broaching or burnishing. These are suitable for all bushing applications in which the tolerable wear will not exceed the thickness of the lining material.

Plain bearing lubrication

The requirements of a lubricant can be summarized as follows:

1) To support the bearing when static and under all speed and load conditions.
2) To have a low coefficient of friction.
3) To be non-corrosive to the materials used in the bearing.
4) To maintain viscosity over the operating range of temperature.
5) Able to provide an effective bearing seal.
6) Have the ability to adhere as a film to the bearing.
7) Be able to conduct heat rapidly.

No single lubricant can satisfy all of these properties and the design of the equipment will determine which aspect needs priority before a choice from available types can be made.

Plain bearing materials

The application of the bearing, the bearing material and the lubricant used are all interdependent, but four basic requirements are necessary for the material:

1) Strong enough to resist failure by fatigue or overheating.
2) Good wear resistance and running properties.
3) Good lubricant retention.
4) High corrosion resistance which may arise due to temperature, the environment and lubricants used.

A wide range of materials consists of metallic, metallic backings with various bearing surfaces, reinforced synthetic resin, graphitic and sintered metallic. Various surface treatments are also available to improve wear resistance and reduce friction.

Whitemetals

These are a large range of either lead base or tin base alloys and are covered by British Standards. Antimony is used as a hardening agent since tin and lead are soft. Whitemetal is a low melting point alloy which is compatible with virtually any type of mating surface. Bearing materials should not be subject to corrosion due to water or the products of oil oxidation and the resistance of tin base whitemetals is high but lead base alloys are susceptible to acidic corrosion from oil oxidation products. Whitemetals are nearly always lubricated under pressure. Loss of lubricant for a short period may cause the bearing to soften and 'wipe'. It loses its compressive strength at elevated temperatures.

Other bearing materials

Other materials for plain bearings include copper lead alloys, lead bronzes, tin bronzes (phosphor bronze), gun metals and aluminium base alloys.

Before concluding this section it should be stated that metallic porous metal bearings are widely used which are manufactured by powder metallurgy where very fine metal powders are mixed and compressed in moulds to the correct form and sintered at high temperature in a reducing atmosphere. The product is in effect a metal sponge which can be impregnated with lubricating oil. The porosity depends on the initial compression and these products are designed for suitable applications where high volume is required. Self lubricating materials are also available in tube and bar form for individual manufacture.

Figs. 28.1 to 28.9 show a selection of different types of bearing from the range manufactured by The Glacier Metal Company Limited. It is generally the case that for small quantities a design, for economic reasons, should incorporate a standard bearing as first choice if possible. Bearing manufacturers employ applications engineers to advise and ensure the correct application and use of their products.

Conventional representation of bearings

A conventional representation is given in BS 308 for a bearing on a drawing and this is shown in Fig 28.27. together with other drawing conventions. The symbol is the same for all types of bearing.

Structural bearings

Plain bearings are also used in structural work and bridge construction to permit expansion and movement and a selection are shown below with an indication of possible motions. This type of bearing utilises the low friction properties of polytetrafluroethylene (PTFE) and in the applications shown can withstand a maximum live loading up to 45 N/mm^2. Illustrations of bearings in Figs 28.1–28.9 are reproduced by kind permission of the manufacturers, The Glacier Metal Company Limited Argyle House, Joel Street, Northwood Hills, Middx. HA6 1LN.

Fig. 28.1 Selection of standard stock bushes manufactured in lead bronze

Fig. 28.2 Types of groove which can be added for lubrication

Fig. 28.3 Prelubricated bearings have an acetal co-polymer lining. The indentations in the linings provide a series of grease reservoirs

Fig. 28.4 Fully machined components from self lubricating materials produced by powder metallurgy

Fig. 28.5 Components pressed to finished size by powder metallurgy techniques

Fig. 28.6 Dry bearings requiring no lubrication with a PTFE/lead bearing surface

Fig. 28.7 Standard wrapped bushes, steel backed lined with lead bronze

Fig. 28.8 Thick walled bearings are produced as bushes, half bearings and thrust washers in copper and aluminium base alloys, also tin and lead base white metals

Fig. 28.9 Structural plain bearings

Ball and roller bearings

Bearing selection

Each type of bearing has characteristic features which make it particularly suitable for certain applications. However, it is not possible to lay down hard and fast rules for the selection of bearing types since several factors must be considered and assessed relative to each other. The following recommendations will, for a given application, serve to indicate those details of greatest importance in deciding the type of bearing to be used.

Available space

In many instances at least one of the main dimensions of the bearing, usually the bore, is predetermined by the machine design. Deep groove ball bearings are normally selected for small diameter shafts whereas cylindrical roller bearings, spherical roller bearings and deep groove ball bearings can be considered for shafts of large diameter.

If radial space is limited then bearings with small sectional height must be selected, e.g. needle roller assemblies, certain series of deep groove bearings and spherical roller bearings.

Where axial space is limited and particularly narrow bearings are required then some series of deep groove ball bearings and cylindrical roller bearings can be used.

Bearing loads

Magnitude of load—This is normally the most important factor in determining the size of bearing. Generally, roller bearings can carry greater loads than ball bearings of the same external dimensions. Ball bearings are mostly used to carry light and medium loads, whilst roller bearings are often the only choice for heavy loads and large diameter shafts.

Direction of load—Cylindrical roller bearings having one ring without flanges and needle roller bearings can only carry radial loads. Other types of radial bearing can carry both radial and axial loads.

Thrust ball bearings are only suitable for axial loads. Spherical roller thrust bearings, in addition to very heavy axial loads, can also carry a certain amount of simultaneously acting radial load.

A combined load comprises a radial and an axial load acting simultaneously.

The most important feature affecting the ability of a bearing to carry an axial load is its angle of contact. The greater this angle the more suitable is the bearing for axial loading. Refer to maker's catalogue for individual values. Double and single row angular contact ball bearings are mainly used for combined loads.

Self aligning ball bearings and cylindrical roller bearings can also be used to a limited extent. Duplex

bearings and spherical roller thrust bearings should only be considered where axial loads predominate.

Where the axial component constitutes a large proportion of the combined load, a separate thrust bearing can be provided for carrying the axial component independantly of the radial load. In addition to thrust bearings, suitable radial bearings may also be used to carry axial loads only.

Angular misalignment

Where a shaft can be misaligned relative to the housing, bearings capable of accommodating such misalignment are required, namely self aligning ball bearings, spherical roller bearings, spherical roller thrust bearings or spherical plain bearings. Misalignments can, for example, be caused by shaft deflection under load, when the bearings are fitted in housings positioned on separate bases and large distances from one another or, when it is impossible to machine the housing seatings at one setting.

Limiting speeds

The speed of rotation of a rolling bearing is limited by the permissible operating temperature. Bearings with low frictional resistance and correspondingly little internal heat generation are most suitable for high rotational speeds. For radial loads, the highest bearing speeds are obtainable with deep groove ball bearings or cylindrical roller bearings and for combined loads the highest bearing speeds are obtainable with angular contact ball bearings.

Precision

Rolling bearings with a high degree of precision are required for shafts where stringent demands are made on running accuracy, e.g. machine tool spindles and usually for shafts rotating at very high speeds.

Deep groove ball bearings, single row angular contact ball bearings, double row cylindrical roller bearings and angular contact thrust ball bearings are manufactured to high degrees of precision both as regards running accuracy and dimensions. When using high precision rolling bearings, shaft and housings must be machined with corresponding accuracy and be of rigid construction.

Rigidity

Elastic deformation in a loaded rolling bearing is very small and in most instances can be ignored. However the bearing rigidity is of importance in some cases, e.g. for machine tool spindles.

Due to the greater area of contact between the rolling elements and raceways, roller bearings, e.g. cylindrical roller bearings or taper roller bearings,

A

Light and
medium radial
loads

D

Heavy
radial
loads

E

Loads in
restricted
radial space

H

J

Axial loads

C **C** **G**

Radial and axial combined loads, α is the angle of
contact

A **F** **B**

Radial and lighter axial combined loads

D **D** **E**

Axial displacement

B **F**

J

Angular misalignment

A **C** **D** **D**

L

D **G**

Rigidity

A **D** **C**

Limiting speeds

Precision

Fig. 28.10

Ease of mounting with cylindrical bore

Adapter
sleeve
added

Withdrawal
sleeve
added

Ease of mounting with taper bore

Fig. 28.10 (Contd.)

deflect less under load than ball bearings. The rigidity of the bearings can be increased by suitable preloading.

Axial displacement

The normal bearing arrangement consists of a locating (fixed) bearing and a non locating (free) bearing. The non locating bearing can be displaced axially thus preventing cross location, e.g. by shaft expansion or contraction. Cylindrical roller bearings having one ring without flanges or needle roller bearings are particularly suitable for use as free bearings. Their internal design permits axial displacement of the inner and outer rings in both directions. The inner and outer rings can therefore be mounted with interference fits.

Mounting and dismounting

The rings of separable bearings (cylindrical roller bearings, needle roller bearings, taper roller bearings) are fitted separately. Thus, when an interference fit is required for both inner and outer rings or where there is a requirement for frequent mounting and dismounting, they are easier to install than non separable bearings (deep groove ball bearings, angular contact ball bearings, self aligning ball bearings and spherical roller bearings).

It is easy to mount or dismount bearings with taper

bores on tapered seatings or when using adapter withdrawal sleeves on cylindrical shaft seatings. Fig. 28.10 gives a simplified guide showing the suitability of the more popular types of bearing for particular applications. The type of bearing indicated for each of the features should be considered as a first choice, but not necessarily the only choice. The bearings listed are described later.

A—Deep groove ball bearing,
B—Self aligning ball bearing,
C—Angular contact ball bearing,
D—Cylindrical roller bearing,
E—Needle roller bearing,
F—Spherical roller bearing,
G—Taper roller bearing,
H—Thrust ball bearing,
J—Spherical roller thrust bearing,
K—Spherical plain bearing,
L—Double row angular contact thrust bearing.

Deep groove ball bearings (Fig. 28.11)

Deep groove ball bearings are available in both single row and double row designs. Single row ball bearings are the most popular of all rolling bearings. They are a simple design, non separable, suitable for high speed operation and require little attention in service. The deep grooves enable axial loads to be carried in either direction. Bearings are available with shields and seals and can be supplied with the correct quantity of lithium base grease and used in operating temperatures between −30° and +110°C. Special bearings operate over a wider range. Relubrication in service is not required. Shielded and sealed bearings are primarily intended for applications where the inner ring rotates. In cases where the outer ring rotates, there is a risk that lubricant will be lost and the manufacturer should be consulted.

Snap rings, fitted to bearings with snap ring grooves provide a simple means of location.

Deep groove ball bearings have very limited ability to accommodate errors of alignment.

Fig. 28.11 Single and double row deep groove ball bearings

Self-aligning ball bearings (Fig .28.12)

Self-aligning ball bearings have two rows of balls and a common sphered raceway in the outer ring and this feature gives the bearing its self aligning property which permits a minor angular displacement of the shaft relative to the housing. These bearings are partic-

Fig. 28.12 Self aligning ball bearings with cylindrical bore

ularly suitable for applications where misalignment can arise from errors in mounting or shaft deflection. A variety of designs are available with cylindrical and taper bores, with seals and adapter sleeves and extended inner rings.

Angular contact ball bearings (Fig. 28.13)

In angular contact ball bearings the line of action of the load, at the contacts between balls and raceways, forms an angle with the bearing axis. The inner and outer rings are offset to each other and the bearings are particularly suitable for carrying combined radial and axial loads. The single row bearing is of non-separable design, suitable for high speeds and carries an axial load in one direction only. A bearing is usually arranged so that it can be adjusted against a second bearing.

A double row angular contact bearing has similar characteristics to two single bearings arranged back to back. Its width is less than two single bearings and can carry an axial load in either direction. These bearings are used for very accurate applications such as the shafts in process pumps.

Fig. 28.13 Single and double row angular contact ball bearings

Cylindrical roller bearings (Fig. 28.14)

In cylindrical roller bearings the rollers are guided between integral flanges on one of the bearing rings. The flanged ring and rollers are held together by the cage to form an assembly which can be removed from the other ring. This separable feature of the bearing design facilitates mounting and dismounting, particularly where, because of loading conditions, interference fits for both rings are necessary. Single and double row bearings are available for heavy loads, high speeds and rigidity. Typical applications are for machine tools and heavy electric motors.

Needle roller bearings (Fig. 28.15)

The chief characteristic of needle roller bearings is that they incorporate cylindrical rollers with a small

Fig. 28.14 Single and double row cylindrical roller bearings

diameter/length ratio. Because of their low sectional height these bearings are particularly suitable for applications where radial space is limited. Needle roller bearings have a high load carrying capacity in relation to their sectional height.

(a)   *(b)*   *(c)*   *(d)*

Fig. 28.15 Needle roller bearings
(a) With inner ring, (b) needle roller cage assembly, (c) drawn cup needle roller bearings with open ends, (d) drawn cup needle roller bearings with closed end

Spherical roller bearings (Fig. 28.16)

Spherical roller bearings have two rows of rollers which run on a common sphered raceway in the outer ring, the inner ring raceways each being inclined at an angle to the bearing axis. The bearings are self aligning and permit minor angular displacements of the shaft relative to the housing which may occur in mounting or because of shaft deflection under load. Heavy duty types are available for severe operating conditions encountered in vibrating machinery such as soil compactors.

Taper roller bearings (Fig. 28.17)

In a taper roller bearing the line of action of the resultant load through the rollers forms an angle with the bearing axis. Taper roller bearings are therefore particularly suitable for carrying combined radial and

Fig. 28.16 Spherical roller bearing

axial loads. The bearings are of separable design, i.e. the outer ring (cup) and the inner ring with cage and roller assembly (cone) may be mounted separately.

Single row taper roller bearings can carry axial loads in one direction only. A radial load imposed on the bearing gives rise to an induced axial load which must be counteracted and the bearing is therefore generally adjusted against a second bearing.

Two and four row taper roller bearings are also made for applications such as rolling mills.

Fig. 28.17

Thrust ball bearings (Fig. 28.18)

Thrust ball bearings are designed to accommodate axial loads. They are not suitable for radial loads. To prevent sliding at the ball to raceway contacts, caused by centrifugal forces and gyratory moments, thrust ball bearings must be subjected to a certain minimum axial load. The bearings are of separable design and the housing and shaft washers may be mounted independently.

Fig. 28.18   Single row thrust ball bearing

Spherical roller thrust bearings (Fig. 28.19)

In spherical roller thrust bearings the line of action of the load at the contacts between the raceways and the rollers forms an angle with the bearing axis, and this

Fig. 28.19   Spherical roller thrust bearing

makes them suitable for carrying a radial load. This radial load must not exceed 55% of the simultaneous acting axial load. The sphered raceway of the housing washer provides a self aligning feature which permits, within certain limits, angular displacement of the shaft relative to the housing.

Application of bearings

A rotating machine element, e.g. the shaft, generally requires two bearings to support and locate it radially and axially relative to the stationary part of the machine, e.g. the housing. Normally, only one of the bearings (the locating bearing) is used to fix the position of the shaft axially, whilst the other bearing (the non-locating bearing) is free to move axially.

Axial location of the shaft is necessary in both directions and the locating bearing must be axially secured on the shaft and in the housing to limit lateral movement. In addition to locating the shaft axially the locating bearing is also generally required to provide radial support and bearings which are able to carry combined loads are then necessary, e.g. deep groove ball bearings, spherical roller bearings and double row or paired single row angular contact ball bearings. A combined bearing arrangement, with radial and axial location provided by separate bearings can also be used, e.g. a cylindrical roller bearing mounted alongside a four-point contact ball bearing or a thrust bearing having radial freedom in the housing.

To avoid cross location of the bearings the non-locating bearing, which provides only radial support, must be capable of accommodating the axial displacements which arise from the differential thermal expansion of the shaft and housings. The axial displacements must be compensated for either within the bearing itself, or between the bearing and its seating on the shaft, or in the housing.

Typical examples of locating and non-locating bearings are shown on the applications in Fig. 28.20.

To prevent roll or creep it is important to maintain the correct fits between the bearings and seatings. Inadequate fits can result in damage to both the bearings and associated components. Normally, the only way to prevent movement at the bearing seatings is to provide a sufficient degree of interference for the bearing rings. Interference fits provide a further advantage in that relatively thin section bearing rings are properly supported around their circumference to give a correct load distribution and allow the load carrying ability of the bearing to be fully utilised. However, where there is a requirement for easy mounting and dismounting of the bearing, or where a non-locating bearing must have freedom of movement axially on its seating, interference fits may not be possible.

Locating   Non-locating   Locating   Non-locating

Locating   Non-locating   Both bearings located

Fig. 28.20

Bearings with cylindrical bore

The most important factors to be considered when selecting bearing fits are as follows:

Conditions of rotation—The conditions of rotation refer to the direction of the load in relation to the bearing rings.

If the bearing ring rotates and the load is stationary, or if the ring is stationary and the load rotates so that all points on the raceway are loaded in the course of one revolution, the load on the ring is defined as a 'rotating load'. Heavy oscillating loads such as apply to the outer rings of connecting rod bearings are generally considered as rotating loads.

If the bearing ring is stationary and the load is also stationary, or if the ring and load rotate at the same speed so that the load is always directed towards the same point on the raceway, the load on the ring is defined as a 'stationary load'.

Variable external loading, shock loading, vibrations and out of balance forces in high speed machines, giving rise to changes in the direction of the load which cannot be accurately established, are classified under the term 'direction of load indeterminate'.

A bearing ring subjected to a rotating load will creep on its seating if mounted with a clearance fit, and wear of the contacting surfaces will occur (fretting corrosion). To prevent this an interference fit should be used. The degree of interference required is dictated by the operating conditions referred to below in the notes on internal clearance and temperature conditions.

A bearing ring subjected to a stationary load will not normally creep on its seating and an interference fit is not therefore necessary unless dictated by other requirements of the application.

When the direction of loading is indeterminate, and particularly where heavy loading is involved, it is desirable that both rings have an interference fit. For the inner ring the fit recommended for a rotating inner ring is normally used. However, when the outer ring must be axially free in its housing or if the loading is not heavy a somewhat looser fit than that recommended for rotating loads may be used.

Magnitude of the load—The load on a bearing inner ring causes it to expand resulting in an easing of the fit on the seating; under the influence of a rotating load, creep may then develop. The amount of interference between the ring and its seating must therefore be related to the magnitude of the load: the heavier the load the greater the interference required.

Internal clearance—When bearing rings are mounted with an interference fit the bearing radial internal clearance is reduced because of the expansion of the inner ring and/or contraction of the outer ring. A certain minimum clearance should however remain. The initial clearance and permissible reduction depend on the type and size of bearing. The reduction in clearance due to the interference fit can be such that bearings with radial internal clearance greater than normal may be necessary.

Temperature conditions—In service, the bearing rings normally reach a higher temperature than the component parts to which they are fitted. This can result in an easing of the fit of the inner ring on its seating or alternatively the outer ring may expand and take up its clearance in the housing thereby limiting its axial freedom. Temperature differentials and the direction of heat flow must therefore be carefully considered in selecting fits.

Requirements regarding running accuracy—Where

bearings are required to have a high degree of running accuracy, elastic deformation and vibration must be minimised and clearance fits avoided. Bearing seatings on shafts should be at least to tolerance IT5 and housing seatings to tolerance IT6. Accuracy of form (ovality and taper) is also very important and deviations from true form should be as small as possible.

Design and material of shaft and housing—The fit of the bearing ring on its seating must not lead to uneven distortion (out of round) of the bearing ring, which may for example be caused by surface irregularities of the seating. Split housings are not suitable when outer rings are to have an interference fit and the limits of tolerance selected should not give a tighter fit than that obtained when tolerance groups H or J apply. To ensure adequate support for bearing rings mounted in thin walled housings, light alloy housings or on hollow shafts, heavier interference fits must be used than would normally be selected for thick walled steel or cast iron housings or solid shafts.

Ease of mounting and dismounting—Bearings having clearance fits are preferred for many applications to facilitate installation and removal. When operating conditions necessitate the use of interference fits and ease of mounting and dismounting is also essential, separate bearings or bearings having a tapered bore and an adapter or withdrawal sleeve can often provide a solution.

Displacement of a non-locating bearing—When a non separable bearing is used at the non-locating position, it is necessary that under all conditions of operation one of the rings is free to move axially. This is ensured by using a clearance fit for that ring which carries a stationary load. Where for example, light alloy housings are used, it may sometimes be necessary to fit a hardened intermediate bush between the outer ring and the housing. If certain types of cylindrical roller bearings, or needle roller bearings are used at the non-locating position, then both inner and outer rings can be mounted with an interference fit.

Bearings with tapered bore

Bearings with a tapered bore are often used to facilitate mounting and dismounting and in some cases this type of bearing may be considered essential to the application. They can be mounted either directly on to a tapered shaft, or by means of an externally tapered sleeve on to a cylindrical shaft.

The axial displacement of a bearing on its tapered seating determines the fit of the inner ring and special instructions relating to the reduction of clearance of bearings with a tapered bore must be observed. The fit of the outer ring in the housing is the same as that for bearings having a cylindrical bore. Adapter and withdrawal sleeves allow greater shaft tolerances to be used (h9 or h10). Errors of form (ovality and taper) of the shaft seating must, however, still be closely controlled (tolerance IT5 or IT7).

Fits and tolerances

Tolerances for the bore and outside diameter of metric rolling bearings are internationally standardized. The desired fits are achieved by selecting suitable tolerances for the shaft and housing using the ISO tolerance system incorporated in BS 4500.

For any particular bearing, the manufacturer's catalogue should be consulted with regard to recommended fits because these must be related to the actual size of the bearings supplied.

Axial location of bearings

Interference fits in general only provide sufficient resistance to axial movement of a bearing on its seating when no axial forces are to be transmitted and the only requirement is that lateral movement of the ring should be prevented. Positive axial location or locking is necessary in all other cases. To prevent axial movement in either direction of a locating bearing it must be located at both sides. When non separable bearings are used as non locating bearings, only one ring, that having the tighter fit, is axially located, the other ring must be free to move axially in relation to the shaft or housing.

Where the bearings are arranged so that axial location of the shaft is given by each bearing in one direction only it is sufficient for the rings to be located at one side only.

Methods of location (Fig. 28.21)

Bearings having interference fits are generally mounted against a shoulder on the shaft or in the housing. The inner ring is normally secured in place by means of a locknut and locking washer (a), or by an end plate attached by set screws to the shaft end (b). The outer ring is normally retained by the housing end cover (c), but a threaded ring screwed into the housing bore is sometimes used (d).

Instead of shaft or housing abutment shoulders, it is frequently convenient to use spacing sleeves or collars between the bearing rings (e), or a bearing ring and the adjacent component, e.g. a gear (f). On shafts, location can also be achieved using a split collar which seats in a groove in the shaft and is retained by either a solid outer ring which can be slid over it, or by the inner ring of the bearing itself.

Axial location of rolling bearings by means of snap rings can save space, assist rapid mounting and dismounting and simplify machining of shaft and housings. An abutment collar should be inserted between the snap ring and the bearing if heavy loads have to be carried, in order that the snap ring is not subjected to large bending moments across its section. If required, the axial clearance, which is generally present between the snap ring and the snap ring groove can be reduced by selecting an abutment collar of suitable width or by

(a) *(b)*

(c) *(d)*

(e) *(f)*

(g) *(h)*

(j) *(k)*

Fig. 28.21 Bearing location methods

using shims. Deep groove ball bearings with a snap ring groove in the outer ring and fitted with a snap ring sometimes provide a simplified and compact housing arrangement.

Bearings with a tapered bore mounted directly on tapered shafts are usually retained by a locknut on the shaft (g), or the locknut may be screwed on to an externally threaded split ring inserted into a groove in the shaft (h). With adapter sleeve mounting, the lock-

nut positions the bearing relative to the sleeve (j). When bearings with an adaptor sleeve are mounted on shafts without an abutment shoulder, the axial load which can be applied depends on the resulting friction between shaft and sleeve. When bearings with a tapered bore are mounted on withdrawal sleeves the inner ring of the bearing must be mounted against an abutment (k). A suitable abutment can be provided by a collar which can frequently serve as part of a labyrinth seal. The withdrawal sleeve must be secured in position either by means of a locknut or an end plate and set screws.

Seals

Bearings must be protected by suitable seals against the entry of moisture and other contaminants and to prevent the loss of lubricant. The effectiveness of the sealing can have a decisive effect on the life of a bearing.

Many factors must be considered when deciding on the best sealing arrangements for a given bearing application, e.g. the type of lubricant (oil or grease), peripheral speed at the sealing surface, misalignment of the shaft, available space, friction of the seal and resultant temperature rise and cost. Two basic designs are normally used for rolling bearings.

Non-rubbing seals (Fig. 28.22)

Non-rubbing seals depend for their effectiveness on the sealing efficiency of narrow gaps, which may be arranged axially, radially or combined to form a labyrinth. This type of seal has negligible friction and wear and is not easily damaged. It is particularly suitable for high speeds and temperatures.

This simple gap type seal which is sufficient for machines in a dry, dust free atmosphere comprises a small radial gap formed between the shaft and housing (a). Its effectiveness can be improved by providing one or more grooves in the bore of the housing cover (b). The grease emerging through the gap fills the grooves and helps to prevent the entry of contaminants. With oil lubrication and horizontal shafts, right or left hand helical grooves can be provided in the shaft or the seal bore (c). These serve to return any oil which may tend to leak from the housing. However, with this arrangement it is essential that the direction of rotation does not vary.

Single or multiple labyrinths give appreciably more effective sealing than gap seals but they are generally more expensive to manufacture. They are chiefly used with grease lubrication. Their effectiveness can be still further improved by providing a grease duct connecting with the labyrinth passage and periodically pumping in a quantity of water insoluble grease, e.g. a calcium soap base grease. In solid housings the tongues of the labyrinth seal are arranged axially (d), and in split housings, radially (e). The radial clearance between the shaft and the housing seal components is not affected by axial displacement of the shaft during running and can be made very small. If angular misalignment of the shaft relative to the housing has to be accommodated, labyrinths of the form shown at (f) are normally used.

An inexpensive and effective labyrinth seal can be made using pressed steel sealing washers (g). The effectiveness of this type of seal increases in direct proportion to the number of washers used. To increase the sealing efficiency of non rubbing seals, the shaft can be fitted with rotating discs (h) and in case of oil lubrication, flinger rings (i) are often used. The oil flung from the ring is collected in a channel in the housing wall and returned to the sump through suitable ducts.

Fig. 28.22

Rubbing seals (Fig. 28.23)

Rubbing seals rely for their effectiveness essentially on the elasticity of the material exerting and maintaining a certain pressure at the sealing surface. The choice of seal and the required quality of the sealing surface depend on the peripheral speed.

Felt washers (a) are mainly used with grease lubrication, e.g. in plummer blocks. They provide a simple seal suitable for peripheral speeds up to 4 m/s and temperatures of about 100°C. The effectiveness of the seal is considerably improved if the felt washer is supplemented by a simple labyrinth ring (b). The felt washers or strips should be soaked in oil at about 80°C before assembly.

Where greater demands are made on the effectiveness of the rubbing seal, particularly for oil lubricated bearings, lip seals are often used in preference to felt seals. A wide range of proprietary lip type seals is available in the form of ready to instal units comprising a seal of synthetic rubber or plastics material normally enclosed in a sheet metal casing. They are suitable for higher peripheral speeds than felt washers. As a general guide at peripheral speeds of over 4 m/s the sealing surface should be ground, and above 8 m/s hardened or hard chrome-plated and fine ground or polished if possible. If the main requirement is to prevent leakage of lubricant from the bearing then the lip should face inwards (c). If the main purpose is to prevent the entry of dirt, then the lip should face outwards (d).

The V-ring seal (e) can be used for grease or oil lubricated bearing arrangements. It comprises a rubber

Fig. 28.23

ring with a hinged rubber lip which is pressed axially against the sealing surface. It is easy to fit, can accommodate fairly large angular misalignments of the shaft relative to the housing at slow speeds, and in certain circumstances is suitable for high speeds. The effectiveness of the seal owes much to the fact that dirt and liquids tend to be flung off by the rotating seal. The V ring seal is normally fitted on the inside rotating seal. The V ring seal is therefore normally fitted on the outside of the housing when grease lubrication is used and on the inside with oil lubrication.

Spring steel sealing washers provide a cheap and space saving seal, especially for grease lubricated deep groove ball bearings. They can either be clamped against the outer ring (f) or against the inner ring and are designed so that the sealing face is constrained to press against the face of the other bearing ring.

Combined seals

In difficult operating conditions and where severe demands are placed on sealing, e.g. large amounts of dirt or water, rubbing and non rubbing seals are often combined. In such cases the non rubbing seals (labyrinths, flinger rings, etc.) are arranged to supplement the rubber seals and protect them from wear.

Sealed and shielded bearings

Simple space saving arrangements can be achieved by using bearings incorporating seals or shields at one or both sides which are supplied lubricated with the correct quantity of grease. Relubrication is not normally required and they are primarily intended for applications where sealing is otherwise inadequate or where it cannot be provided for reasons of space.

Lubrication

Grease lubrication is generally used where ball and roller bearings operate at normal speeds, temperature and loading conditions. Grease has certain advantages by comparison with oil: it is more easily retained in the bearing housing and assists in sealing against the entry of dirt and moisture.

In general the free space in the bearing and housing should only be partly filled with grease (30 to 50%). Overfilling causes rapid temperature rise particularly if speeds are high. Manufacturers supply details regarding suitable weights of grease for particular bearings.

Bearings can be supplied which are sealed after prepacking with the correct type and quantity of grease. Where relubrication is more frequent, provision must be made by fitting grease nipples to the housing. Grease will then be applied by a grease gun and a lubrication duct should feed the grease adjacent to the outer ring raceway or between the rolling elements. Examples are shown in Fig. 28.24.

Fig. 28.24

Oil lubrication

Oil lubrication is generally used where high speeds or operating temperatures prohibit the use of grease, when it is necessary to transfer frictional heat or other applied heat away from the bearing, or when the adjacent machine parts, e.g. gears, are oil lubricated.

Oil bath lubrication is only suitable for slow speeds. The oil is picked up by rotating bearing elements and after circulating through the bearing drains back to the oil bath. When the bearing is stationary the oil should be at a level slightly below the centre of the lowest ball or roller. An application is shown in Fig. 28.25. At high speeds it is important that sufficient oil reaches the bearing to dissipate the heat generated by friction and oil jets provide an effective method (Fig. 28.26).

Fig. 28.25

Fig. 28.26

The illustrations in this section (Figs 28.10 to 28.26) are reproduced by kind permission of SKF (U.K.) Limited, Bradbourne Drive, Tilbrook, Milton Keynes, MK7 8BJ.

Trouble-Free Bearing Operation

When bearings fail, they can bring equipment to an unscheduled halt. Every hour of downtime due to premature bearing failure can result in costly lost production in a capital intensive industry. Substantial investment in research and development has resulted in the manufacture of bearings of the highest quality. Quality alone cannot guarantee trouble-free bearing operation since other factors may affect life span including the following:

1 *Operating environment*   Machinery must be kept in peak operating condition. Bearings should be properly aligned and protected from extreme temperatures, moisture and contaminants.
2 *Proper installation*   Knowledge of the proper installation techniques and tools is required to ensure that the bearings are not damaged.
3 *Proper maintenance*   Following lubrication and maintenance schedules using recommended materials and time intervals is essential. A familiarity with operating procedures, basic trouble shooting, condition monitoring and vibration analysis is also desirable.

However, bearing manufacturers do have a full line of products and services to make installation and maintenance easy to perform and should be consulted. This will certainly contribute to long bearing life and ensure cost effective operation.

BS convention

A general convention for rolling bearings in section does not differentiate between each type or detail and is given in Fig. 28.27.

The boundary of the bearing is indicated on the drawing with diagonal lines added. The drawing parts list or schedule will then provide details of the exact bearing specification.

Ball and roller bearings

BS convention

Fig. 28.27

Chapter 29
Engineering adhesives

The use of adhesives is now a well established practice in manufacturing. New materials and production processes have considerably increased the options available to the engineering designer. Adhesive bonding is a proven cost effective manufacturing method and can be used with confidence. A basic principle is however that joints should be designed with this method of production in mind when the product is in the early stages of development.

The following are some advantages of using adhesives:

a) Stress concentrations present in bolted, riveted or spot welded joints are avoided.
b) The distribution of stresses achieved by adhesive bonding permits a reduction in weight and cost. Especially relevant with fragile materials and lightweight structures. Joint strength and fatigue properties are improved.
c) Production costs are reduced due to the elimination of drilled holes and other machining operations. Labour costs are reduced with automated assembly work.
d) Structures are generally stiffer despite weight reduction since the bonding covers the whole area of the join. Rivets, screws and spot welds pin the surfaces together only at localised points. Loading may be increased before buckling occurs.
e) Gap filling properties. Certain adhesives are gap filling, and this makes possible the continuous joining of materials where the gap along the joint is of irregular width.
f) Delicate or brittle materials such as metal foils or ceramics are readily bonded.
g) High strength bonds can be formed at room temperature with minimal pressure by using cold-setting adhesives.
h) The film formed by the adhesive resists corrosion, can form a leak-proof seal and insulate dissimilar metals against electrochemical action.

Designing for adhesives

For the best possible performance, joints should be specifically designed for adhesive bonding. Follow this principle and much better joints will be achieved than if bonding is adopted as a substitute for welding in a joint designed for that purpose. Bond stresses, materials,

type of adhesive, surface preparations, method of application and production requirements can then all be considered in relation to each other at the outset. The designer should consider especially the effect of shear, tension, cleavage and peel stresses upon the joint. Bonded joints perform best under conditions of tension (pure), compression or shear loading; less well under cleavage; and relatively poorly under peel loading. The loading condition are shown in Fig. 29.1.

Designing a joint to take pure tensile or compressive stresses is normally impracticable with sheet materials, so all joints in sheet materials should be designed so that the main loading is in shear. Joints between massive parts perform well in tension or compression loading, provided this is uniform–a side load may set up excessive cleavage stresses in a tension-loaded bond. (Fig. 29.1(d)). Cleavage loading will concentrate stress at one side of the joint. Bond area may have to be increased to withstand this load so the joint will not prove so economical in terms of material and/or adhesives as joints designed for shear and tension stresses. Peel strength is usually the weakest property of a joint. A wide joint will be necessary to withstand peel stresses, plus the use of an adhesive with high peel strength.

For an adhesive to be used, a joint must allow the easy application of the adhesive, must allow for the adhesive to cure fully, and must be designed to give

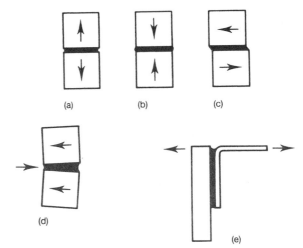

Fig. 29.1 Loading conditions.
(a) Tension (b) Compression
(c) Shear (d) Cleavage
(e) Peel

uniform stress. Even in a simple face-to-face joint it must be possible to apply adhesive to one surface and for it to remain there until the two parts are brought together and after that until curing takes place.

These requirements highlight the need for a choice of thin, thick or thixotropic adhesives. Design details which may also be significant include removal of sharp edges and substitution of a bevel or radius.

The bond line

The gap between the parts, and therefore the thickness of the adhesive film, has an important bearing on the characteristics of the joint. In terms of simple strength a thick bond line will generally be a weakening feature, since the mechanical strength of the unsupported resin film is likely to be less than that of the substrates.

A thick bond line can however confer advantages. The adhesive is generally more flexible than the adherents or substrates. This is particularly so in most engineering applications where metals or other rigid materials can be bonded. Because of this, a thick bond line can offer a capacity to absorb some impact energy, thus increasing the strength of the bond under this type of loading.

Consideration of bond line thickness leads immediately to the question of environmental resistance.

Adhesive bonds will always be susceptible to environmental attack and it is essential that any such attack should not reduce the strength of the bond to an unacceptable level. The most important factor here is the correct choice of adhesive, but design of the joint can make a significant difference. Thus a thick bond line offers a ready path for access by moisture or other solvents which might be able to diffuse through the cured adhesive.

Typical bonded joints

Fig. 29.2. shows a range of bonded joints and possible modifications which can be made to reduce or eliminate the effect of cleavage and peel stresses.

The following notes should be regarded as of a general nature.

a) Avoid butt joints if bond area is small.
b) Scarfed joint provides increased bonding area.
c) Simple lap joint with in-line forces.
d) Alternative lap joint with offset loading.
e) Tapered lap joint.
f) Bracket bonded to a fixed surface where peel is likely.
g) Repositioned bracket strengthens joint.
h) & (j) Cleavage loading eliminated by the addition of a component in shear.
k) & (l) Simple improvement for safety.
m) & (n) Increase in bond area reinforces corner joint.

Quite obviously practical considerations involve a study of the forces applicable and acceptable appearance of the finished assembly.

Fig. 29.4. shows two tubular applications.

In (a), a cylindrical plug is used to join two tubes in a structure. An example of a tapered tubular joint is given in (b). The taper ensures that the adhesive is not pushed out of the assembly.

The joint permits a long bond line and does not impede fluid flow.

A selection of bonded stiffeners are shown in Fig. 29.5. These can be used to reduce vibration and deflection of thin sheet materials. When the flanges on the stiffened sections do deflect with the sheet, little difficulty from peel results due to the area of the bond. Corrugated backings can provide complete flatness over the entire area. If a corrugated insert is sandwiched between two flat sheets (the second sheet is indicated by dotted lines) as indicated in example (a) then a structure of lightweight and high strength can be manufactured from adhesive bonding. There are many aircraft applications. Standard strip, angles, tee sections and formed channels are used in structural engineering.

The types of adhesive which cover the vast majority of engineering assembly applications come from the following categories.

1 *Epoxies*   Two components are mixed in equal proportions. The adhesive and the hardener begin to cure immediately and have a usable 'pot life'. After this time the bond becomes less effective. Often used for DIY repairs. Industry uses an alternative type of epoxy which incorporates rubber of low molecular weight and is called a toughened adhesive. It has greater resistance to impact forces and peel.

This is a single component epoxy which is hardened by heat curing while the parts being bonded are clamped.

Used to bond composite materials, tubular frames and in the manufacture of components for double glazing assemblies.

2 *Acrylic adhesives*   Four basic types:
a) Toughened acrylics. These are two-part systems where a hardener and an adhesive are applied to the two surfaces being joined and the assembly of the joint automatically mixes them. Can be used on oily steel. Will bond glass into metal frames. Also used in railway carriage interior panels.
b) Cyanoacrylate adhesives polymerise (solidify) by a chemical reaction which is usually initiated by atmospheric moisture, present as traces of water on the surfaces to be joined. Successful bonding depends upon ambient humidity, the type of material being bonded, the choice of adhesive, and the nature of the surface.

'Instant adhesives' and 'Superglues' are in this range of products.

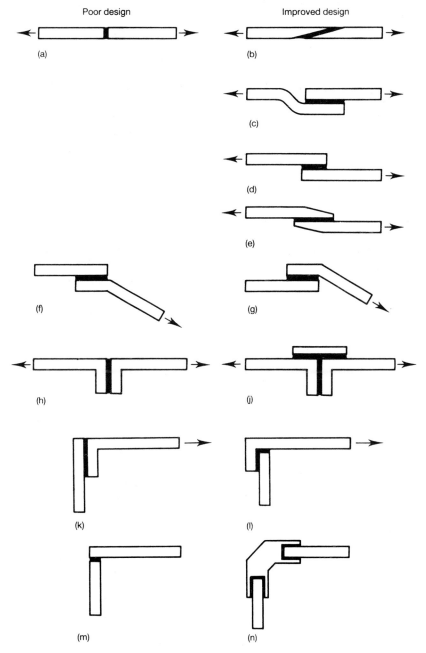

Poor design   Improved design

(a)   (b)

(c)

(d)

(e)

(f)   (g)

(h)   (j)

(k)   (l)

(m)   (n)

Fig. 29.2 Typical bonded joints

Fig. 29.3 Where slotted joints are used, tapering removes the high stress concentrations caused by abrupt changes in section. Example gives a possible modification to Fig. 29.2(n)

(a)

(b)

Fig. 29.4

(a)

(b)

(c)

(d)

(e)

Fig. 29.5

c) Anaerobic adhesives automatically harden in the absence of air and are used mainly in rigid metallic joints. Many applications follow. These products are manufactured normally as single component materials.

d) UV curing acrylics are single component adhesives where cure is effected rapidly by placing the assembly under an ultra violet lamp.

These adhesives are applied in the manufacture of printed circuit boards for surface sealing.

3 Hot melt adhesives are available in rod, sheet and powder forms. A convenient method of assembling small components which are lightly loaded. A heating gun raises the temperature of the rod and the adhesive is applied to one component. On cooling, the adhesive solidifies and the two surfaces are bonded together. These adhesives are also used in packaging equipment.

4 Solvent based contact adhesives. Here the adhesive is applied in a solvent solution to the two surfaces. The solvent evaporates leaving a tacky film and the surfaces are brought together. Applications include laminated sheet fixings in furniture manufacture.

A considerable range of options is available to the designer in the choice of suitable types of adhesive.

Precision measuring and dispensing is necessary so that the required volume, in the defined position, is applied at a given time and with consistently repeatable accuracy on a production line.

In the interests of satisfactory selection and operation, it is recommended that the manufacturer should be consulted to ensure that all technical considerations have been included in a proposed scheme.

Engineering applications

The following examples show varied uses of engineering adhesives in industry.

Locking screw threads The liquid is applied to the cleaned thread of a bolt or stud. When the nut is tightened the liquid fills the gaps between mating threads and hardens to form a tough plastic joint which is shock, vibration, corrosion and leak proof. The joint will remain in place until it needs to be undone again using normal hand tools.

Fig. 29.6 Thread locking

Threadsealing pipe fittings The sealant is applied to the clean thread and screwed together as normal. The sealant will not creep or shrink and gives a constant and reliable seal. There is no need to wrench tight and the fitting can be positioned as required.

Fig. 29.7 Thread sealing

(a) Hydraulic sealant for fine threads in pneumatic and hydraulic systems—particularly those subject to vibration

(b) Pipe sealant used to seal coarse threads of pipes and fittings up to 75 mm outside diameter

Retaining Traditional retaining methods using screws, splines, pins, keys and press fits etc, do not necessarily seal joints and eliminate the possibility of corrosion. Local stress concentrations may cause cracking. Retaining adhesives can be used to augment these methods. Often, a redesign will give a replacement with substantial cost savings.

These adhesives are supplied in various strengths:

a) High shear strength adhesives in association with press fits can provide added rigidity

Engineering applications 257

Fig. 29.8   Retaining

b) Maximum strength retainers are used on parts which generally do not need to be taken apart.
c) Medium strength adhesives suit parts which need frequent disassembly.

Sealing with anaerobic gaskets   Gaskets are fitted between flanges to provide an effective seal against fluids and gases. It is cheaper to use a gasket than manufacture two perfectly flat mating surfaces with close flatness and surface finish tolerances.

Gaskets can be preformed from materials such as compressed asbestos, paper, fibre or cork. Alternatively, they can be formed where they are required with a liquid.

The principles of liquid gasketing are fundamentally different to preformed gaskets in that they allow metal to metal contact. There are several forms of liquid gasket such as anaerobic, non setting solvent based and moisture curing.

The anaerobic principle   Anaerobic gaskets are available in a range of viscosities from thick liquids to non slump pastes. Each can be applied directly from the original container, or by various application methods such as simple rollers, screen printing and computerised robotics. On assembly, the anaerobic gasket spreads between the flanges and is forced into surface irregularities to provide total contact between the two faces. The product then polymerises at ambient temperature into a tough thermoset plastic.

The strength of joints from anaerobics can be tailored to suit a specific application. Effective cure requires the absence of air and the presence of metal. At room temperature it takes just a few minutes.

Note. Anaerobic gaskets are thermosetting plastics; the temperature range in service can be from $-50°C$ up to $200°C$ at the joint line. They seal against petroleum based fuels and lubricating oils, water/glycol mixtures and many other industrial chemicals. For compatibility of specific chemical environments the designer would be advised to consult the manufacturers.

Although anaerobic gaskets permit metal to metal contact, electrical continuity cannot be assumed.

Fig. 29.9 shows the application of an anaerobic gasket to the backplate of a large diesel engine.

The flow of adhesive to the work surface is regulated by manual control of the air supply to a pneumatic cartridge gun.

It often happens during maintenance work that damaged or scored surfaces are found and a adhesive gasket can save the need and cost of remachining.

Engineering adhesives for sealing flat faces have the following characteristics and applications.

a) They will seal on horizontal, vertical and overhead flanges and accommodate surface irregularities of up to 0.5 mm.
b) Low strength products are available for close fitting surfaces which will be frequently dismantled.
c) In the illustrations below (Fig. 29.10) many of the components are manufactured in aluminium alloys. The structural integrity of an assembly can be enhanced by the use of high shear strength adhesives.

Fig. 29.9

Fig. 29.10

Engineering adhesives for retaining cylindrical assemblies have the following characteristics and applications:

a) The retention of shafts and rotors of electric motors, gears, pulleys, sleeves, bushes and oils seals in housings.
b) The ability to withstand fatigue and augment torsional strength.
c) Suitable for parts that need easy disassembly, such as bearings on shafts and in housings, bushes and journals in soft metals.
d) An oil-tolerant adhesive is available that gives high strength retention of parts 'as received', i.e. no cleaning is needed before assembly. Oil impregnated bushes are retained with this grade. They are manufactured by the sintering process.
e) An adhesive can be recommended for continuous working temperatures up to 175°C. It combines the ability to fill gaps of up to 0.15 mm in diameter with high shear strength and good solvent resistance.

Instant adhesives

As the name suggests, they work in seconds and are ideal for bonding close fitting parts made from a variety of materials. They offer advantages over other forms of joining, such as plastic welding, two-part or heat-cured adhesives and mechanical fasteners. The benefits include faster assembly times, better appearance, less outlay for capital equipment and these adhesives can also be used to repair metal, plastic, rubber or ceramic components which might otherwise be scrapped.

Instant adhesives are available for the following applications:

a) General purpose adhesive for plated metals, composite materials, wood, cork, foam, leather, paper—all surfaces which were once considered 'difficult'—can now be bonded quickly and permanently.
b) A special rubber and plastics adhesive ensures fast-fixturing of elastomers and rubbers, especially EPDM rubber. Bonds polyethylene, polypropylene and polyolefin plastics.
c) A gel-type adhesive can be used for fabrics, paper, phenolic, PVC, neoprene and nitrile rubber and

Fig. 29.11

bond them in 5 seconds; ceramic, leather, and balsa wood in 10; mild steel in 20; ABS and pine in 30. The gel form prevents absorption by porous materials and enables it to be applied to overhead and vertical surfaces without running or dripping.

d) A low odour, low bloom adhesive has been developed where application vapours have been removed with no possibility of contamination. A cosmetically perfect appearance can be obtained. The absence of fumes during application means that it can be safely used close to delicate electrical and electronic assemblies, alongside optics and in unventilated spaces.

e) A black rubber toughened instant adhesive gives superior resistance to peel and shock loads. Tests show bonds on grit blasted mild steel can expect a peel strength of 4 N/mm at full cure.

All adhesives can be applied direct from bottle, tube or standard automatic application equipment on to surfaces which require very little pretreatment.

In most cases, just one drop of instant adhesive is enough to form an extremely strong, virtually unbreakable bond. There is no adhesive mixing, and cure takes place in seconds to give a joint with maximum surface to surface contact.

Four typical production applications are shown in Fig. 29.12.

The illustration in Fig. 29.12(c) shows an operator using a semi automatic dispenser. The bonding product is contained in a bottle pack and dispensing regulated by an electronic timer controlled pinch valve mounted on the unit. The dispense function can be initiated in a variety of ways, including a footswitch. The point of application is controlled by hand.

Structural applications

Structural adhesives are ideal for bonding large areas of sheet materials. They can produce a much better finished appearance to an assembly than, say, rivets, or spot welding or screws. The local stress introduced at each fixing point will be eliminated. Furthermore, adhesives prevent the corrosion problems normally associated with joining dissimilar materials. This is a cost effective method of providing high strength joints.

The authors wish to express their thanks for the assistance given and permission to include examples showing the application of adhesives in this chapter by Loctite UK, a division of Loctite Holdings Ltd, Watchmead, Welwyn Garden City, Hertfordshire AL7 1JB.

(a) Bonding rubber to uPVC double glazing units

(b) Bonding toroid to the PCB from a temperature control unit

(c) Bonding brass to PVC on a connector

(d) Bonding foam rubber to moulded polyurethane grouting tool

Fig. 29.12

Fig. 29.13 A structural adhesive used to bond a stiffener to an aluminium car bonnet. To line up the two parts a purpose made fixture is designed

Fig. 29.14 Mild steel stiffeners are bonded to up and over garage door. Result: rigidity, unblemished exterior surfaces

Chapter 30
Related standards

Standards are essential tools for industry and commerce, influencing every aspect of the industrial process. They provide the basic ingredients for competitive and cost-effective production.

Standards define criteria for materials, products and procedures in precise, authoritative and publicly available documents. They embrace product and performance specifications, codes of practice, management systems, methods of testing, measurement, analysis and sampling, guides and glossaries.

Thus they facilitate design and manufacture;
establish safety criteria;
promote quality with economy;
assist communication and trade; and
inspire confidence in manufacturer and user.

The role of standards in national economic life is expanding. They are increasingly referred to in contracts;
called up in national and Community legislation;
used as a basis for quality management;
required for product certification;
required in public purchasing; and are
used as a marketing tool

The British Standards Institution

The BSI was the worlds first national standards body. There are now over 80 similar organizations worldwide which belong to the International Organization for Standardization (ISO) and the International Electrotechnical Commission (IEC). BSI represents the views of British Industry on these bodies, working towards harmonizing world standards. BSI also presents the British view in discussions on the European Standards to be used in the single market after 1992.

The BSI catalogue lists over 10 000 publications. Each year more than 700 new or revised standards are issued to keep the technical content up to date due to developments in materials, processes and technologies. BSI Quality Assurance is the largest independant certification body in the UK and provides a comprehensive service including certification, registration, assessment and inspection activities.

BSI Testing at Hemel Hempstead offer services which are completely confidential and cover test work in many areas including product safety, medical equipment, motor vehicle safety equipment and calibration work. Tests are undertaken to national or international standards.

The BSI Library holds a collection of 500 000 international and foreign standards, codes of practice, regulations and technical requirements. Most of these are available on loan to UK subscribing members in exchange for library tokens purchased from the BSI Sales Department.

Further information on BSI services can be obtained from:
BSI Enquiry Department,
Linford Wood,
Milton Keynes MK14 6LE
Tel: 0908 221166

Complete sets of British Standards are maintained for reference purposes at many Public, Borough and County Libraries in the UK. Copies will be also available in University and College of Technology Libraries.

The first edition of BS 308 which covered Engineering Drawing Office practice was introduced in 1927. The recommendations on four pages of slightly less than A5 size were accompanied by five pages of illustrations.

The standard covered drawing sizes, number, date, name, scale, projection, examples of linework, sections, screw threads and tapers. Abbreviations to be used on drawings were given for general terms and ferrous, non-ferrous and non-metallic materials.

On the rear cover part of a short note stated that 'The British Engineering Standards Association exists for the purpose of drawing up British Standard Specifications for materials, machinery, or apparatus. It does not, however, embark on such work on its own initiative, but at the specific request of an authoritative body such as a representative trade organization, a technical society, or a Government department, and to fill a recognised want.'

The same basic philosophy exists today, but the standard has been continuously developed, and is now published in three parts:
Part 1. General principles
Part 2. Dimensioning and tolerancing of size
Part 3. Geometrical tolerancing.

The Standards-making process

The BSI Standards function is to draw up voluntary standards in a balanced and transparent manner, to reach agreement among all the many interests concerned, and to promote their adoption.

Standards are drafted by technical committees whose members are nominated by manufacturers, trade and research associations, professional bodies, central and local government, academic bodies, user and consumer groups. BSI arranges the secretariats and takes care to ensure that its committees are representative of the interests involved. Members and Chairmen of committees are funded by their own organizations.

Proposals for new and revised standards come from many sources but the largest proportion is from industry. Each proposal is carefully examined against its contribution to national needs, existing work programmes, the availability of internal and external resources, the availability of an initial draft and the required timescale to publish the standard. If the work is accepted it is allocated to a relevant existing technical committee or a new committee is constituted.

Informed criticism and constructive comment during the committee stage are particularly important for maximum impact on the structure and content of the future standard.

The draft standards are made available for public comment and the committee considers any proposals made at this stage. The standard is adopted when the necessary consensus for its application has been reached.

Strategy, policy, work programmes and resource requirements are formulated and managed by Councils and Policy Committees covering all sectors of industry and commerce.

The following table shows aspects of engineering drawing practice covered by British and International (ISO) Standards. The numbers given merely indicate the Standards in which the subjects are covered and no equivalence is implied.

Subject	BS	ISO
Centre holes: simplified representation	—	6411
Conventional representations	308: Part 1	—
Definitions		
dimensioning and tolerancing	308: Part 2	129
general	308: Part 1	—
Dimensioning and tolerancing		
cones	—	3040
profiles	—	1660
Dimensions	308: Part 2	129
Drawing sheets		
layout	308: Part 1	5457
sizes	3429	5457
Drawing types	308: Part 1	—
Engineering diagrams	5070	—
Folding of drawing prints	1192: Part 1	
Gears	308: Part 1	2203

Subject	BS	ISO
Geometrical tolerancing		
datums	308: Part 3	5459
general	308: Part 3	1101
maximum material condition	308: Part 3	2692
positional tolerancing	308: Part 3	5458
Indication of surface texture	308: Part 2	1302
Item lists	—	7573
Item reference	308: Part 1	6433
Laboratory glassware	2774	6414
Lettering	308: Part 1	3098/1
Lines	308: Part 1	128
Optical elements and systems	4301	—
Pictorial representations	1192: Part 1	5456
Principles of dimensioning	308: Part 2	129
Projection and dimension lines	308: Part 2	129
Projections	308: Part 1	128
Requirements for microcopying	5536	6428
Scales	308: Part 1	5455
Sections	308: Part 1	128
Splines and serrations	308: Part 1	6413
Springs	308: Part 1	2162
Symbols		
geometrical tolerances	308: Part 3	7083
welding	499	2553
and abbreviations	308: Part 1	—
Tapered features	308: Part 2	—
Terminology of dimensions and tolerances	308: Part 2	5463
Threaded parts	308: Part 1	6410
Title block	308: Part 1	7200
Toleranced dimensions	308: Part 2	129: and 406
Tolerancing principles	308: Part 2	8015
Views	308: Part 1	128

International Organization for Standardization (ISO)

ISO (the International Organization for Standardization) is a worldwide federation of national standards bodies, at present comprising 89 members, one in each country. The object of ISO is to promote the development of standardization and related activities in the world with a view to facilitating international exchange of goods and services, and to developing co-operation in the sphere of intellectual, scientific, technological and economic activity. The results of ISO technical work are published as *International Standards*.

The scope of ISO covers standardization in all fields except electrical and electronic engineering standards, which are the responsibility of the International Electrotechnical Commission (IEC).

ISO brings together the interests of producers, users (including consumers), governments and the scientific community, in the preparation of International Standards.

ISO work is carried out through some 2 600 technical bodies. More than 20 000 experts from all parts of the world participate each year in the ISO technical work which, to date, has resulted in the publication of 7 778 ISO standards.

Origin

International standardization started in the electro-technical field some 80 years ago. While some attempts were made in the 1930s to develop International Standards in other technical fields, it was not until ISO was created that an international standards organization devoted to standardization as a whole came into existence.

Following a meeting in London in 1946, delegates from 25 countries decided to create a new international organization 'the object of which would be to facilitate the international coordination and unification of industrial standards'. The new organization, ISO, began to function officially on 23 February 1947.

Members

A *member body* of ISO is the national body 'most representative of standardization in its country'. It follows that only one such body for each country is accepted for membership of ISO. Member bodies are entitled to participate and exercise full voting rights on any technical committee of ISO, are eligible for Council membership and have seats in the General Assembly.

By January 1991, the number of member bodies was 72.

More than 70% of the ISO member bodies are governmental institutions or organizations incorporated by public law. The remainder have close links with the public administration in their own countries.

Basic information on each ISO member body is given in the publication *ISO Member Bodies*.

A *correspondent member* is normally an organization in a developing country which does not yet have its own national standards body. Correspondent members do not take an active part in the technical work, but are entitled to be kept fully informed about the work of interest to them. They may attend the General Assembly as observers. Nearly all the present correspondent members are governmental institutions.

By January 1991, the number of correspondent members was 17.

Technical work

The technical work of ISO is carried out through *technical committees (TC)*. The decision to establish a technical committee is taken by the ISO Council and its scope is approved by the ISO Technical Board on behalf of Council. Within this scope, the committee determines its own programme of work.

Work in the field of information technology is carried out through a joint ISO/IEC technical committee, ISO/IEC JTC 1 *Information technology*, established in 1987 by the ISO and IEC Councils.

Each technical committee may, in turn, establish *sub-committees (SC)* and *working groups (WG)* to cover different aspects of its work.

Each technical committee or sub-committee has a *secretariat*, assigned to an ISO member body: in the case of technical committees, by the Technical Board on behalf of Council, and in the case of sub-committees, by the parent committee. For each working group, a convener is appointed by the parent committee.

By the end of 1990, there were 172 technical committees, 653 sub-committees, 1764 working groups and 10 ad hoc study groups.

A proposal to begin work in a new field of technical activity normally comes from within ISO itself, but it may also originate from some other international organization. Since the resources are limited, priorities must be considered. Therefore, all new proposals are submitted for consideration by the ISO member bodies. If accepted, either the new work will be referred to the appropriate existing technical committee or a new technical committee will be established.

To ensure coordination of work in all matters of common interest, liaisons are established between related technical committees. Each member body interested in a subject for which a technical committee has been authorized has the right to be represented on that committee.

Member bodies which decide to take an active part in the work of a technical committee or sub-committee are designated as *P-members* (participating members) of that committee or sub-committee. They have an obligation to vote and, whenever possible, to attend meetings. One of the P-members is designated to act as the secretariat of the committee or sub-committee.

Member bodies which wish only to be kept informed of the work of a technical committee or sub-committee are registered as *O-members* (observers).

Detailed rules of procedure for the technical work are given in the *IEC/ISO Directives*.

How an International Standard is developed

An International Standard is the result of an agreement between the member bodies of ISO.

An International Standard may be used as such, or may be implemented through incorporation in national standards of different countries.

A first important step toward an International Standard is the *committee draft (CD)*–a document circulated for study within the technical committee or sub-committee.

This document must pass through a number of stages before it can be accepted as an International Standard. This procedure is designed to ensure that the final result is acceptable to as many countries as possible.

When agreement is finally reached within the committee, the committee draft is sent to the Central Secretariat for registration as a *Draft International Standard (DIS)*; the DIS is then circulated to all member bodies for voting. If 75% of the votes cast are in favour of the DIS, it is accepted for publication as an International Standard. Normally the fundamental technical issues are resolved at committee level; however, the member body voting procedure provides assurance that no important objections have been overlooked.

The greater part of the work is done by correspondence, and meetings are convened only when thoroughly justified. Each year some 10 000 working documents are circulated.

Most standards require periodic revision. Several factors combine to render a standard out of date: technological evolution, new methods and materials, new quality and safety requirements. To take account of these factors, ISO has established the general rule that all ISO standards should be reviewed at intervals of not more than five years. On occasion it is necessary to revise a standard earlier.

A list of all ISO standards is given in the *ISO Catalogue*.

External relations

ISO work is of interest to many other international organizations: some of these make a direct technical contribution to the preparation of ISO standards; others, particularly the inter-governmental organizations, contribute to the implementation of ISO standards, for example by utilizing them in the framework of inter-governmental agreements. ISO has adopted arrangements for associating these organizations closely with all stages of the work:

—international organizations may make proposals for the preparation of ISO standards in a new field in the same way as ISO member bodies;
—international organizations may be granted 'liaison status' with technical committees and sub-committees. Liaison status comprises two categories: 'A' (effective contribution to the work) and 'B' (wish to be kept informed only). Liaison A gives the right to submit papers, attend meetings, and participate in discussions;
—international organizations which can make an effective contribution to the implementation of ISO standards are expressly invited to comment on all relevant drafts;
—technical committees are instructed to seek the full and, if possible, formal backing of the main international organizations in liaison for each ISO standard in which these organizations are interested.

Some 450 international organizations have liaison status with ISO; this includes all UN specialized agencies working in similar fields. ISO has consultative status (Category I) with the UN Economic and Social Council (ECOSOC) and equivalent status with nearly all other bodies and specialized agencies in the UN system.

A full list of liaisons between the international organizations and ISO technical committees and sub-committees is contained in the publication *ISO Liaisons*.

ISO also maintains close working relations with regional groups of standards bodies. In practice the members of such regional groups are also members of ISO and the principle is generally accepted that ISO standards are taken as the basis for whatever standards are required to meet the particular needs of a given geographical region.

The address of ISO is:
1, rue de Varembé
Case postale 56
CH-1211 **Genève 20**
Switzerland/Suisse
Telephone: +41 22 749 01 11
Telefax: +41 22 733 34 30
Telex: 41 22 05 iso ch
Telegrams: isorganiz

International Electrotechnical Commission (IEC)

The *International Electrotechnical Commission* came into being in 1906.

ISO and IEC have established a formal agreement, stating the relationship between the two organizations. In accordance with this agreement, ISO and IEC form the specialized system for worldwide standardization.

It is agreed that questions related to international standardization in the electrical and electronic engineering fields are the concern of IEC. Other subject areas are the responsibility of ISO. In matters of international standardization not related to any particular technology ISO undertakes, in consultation with IEC, to safeguard any electrotechnical interests which may be involved. To ensure the necessary technical coordination, ISO and IEC have established a joint ISO/IEC Technical Programming Committee.

The members of the IEC are the *national committees*, one for each country, which are required to be as representative as possible of all electrical interests in the country concerned: manufacturers, users, governmental authorities, teaching and professional bodies. They are composed of representatives of the various organizations which deal with questions of electrical standardization at the national level. Most of them are recognized and supported by their governments.

In the technical work, numerous liaisons have been established between ISO and IEC committees, and a joint ISO/IEC technical committee has been established in the field of information technology.

Further information on the IEC and its national committees can be found in the *IEC Directory*.

The address of IEC is:
3, rue de Varembé
CH-1211 **Genève 20**
Switzerland (Suisse)

Telephone: +41 22 734 01 50
Telefax: +41 22 733 38 43
Telex: 41 41 21 iec ch
Telegrams: inelission-genève

CEN (European Committee for Standardization Comité Européen de Normalisation)

The International and European Connection

Crucial to the success of the UK in world markets is a strong voice in the preparation of international and European standards. These essential documents underpin trade and the greater the influence that can be brought to bear on their content, the more easily UK industry can satisfy overseas markets.

BSI is a founder member of the International Organization for Standardization (ISO), the International Electrotechnical Commission (IEC) and their European counterparts–the European Committee for Standardization (CEN) and the European Committee for Electrotechnical Standardization (CENELEC). The 'gateway' to participation in these bodies is first through representation on BSI's technical committees and then appointment to UK delegations and expert working groups. Seventy per cent of BSI's projects are directly concerned with the production of European and international standards.

An important feature of the European work is that while agreement on technical content in committee is reached by consensus, adoption of the standard is by weighted majority voting. Adopted standards must be implemented as national standards, regardless of the way the national member voted and any conflicting standards have to be withdrawn.

It is essential for BSI, through its committee structure, to maintain a high profile within the technical harmonization programme to ensure the resulting standards fulfil UK needs. This is particularly important when standards are to be linked to European Community Directives for completing the Single Market thereby giving the European standard a legal status as a means of compliance.

Member countries of the European Community are represented by CEN through their own national organisations as follows:

BELGIUM/BELGIQUE (IBN)
Institut belge de normalisation
Av. de la Brabançonne 29
B-1040 Bruxelles

DENMARK/DANEMARK (DS)
Dansk Standardiseringsraad
Aurehøjvej 12
DK-2900 Hellerup

FRANCE (AFNOR)
Association française de normalisation
Tour Europe
Cedex 7
F-92049 Paris La Défense

GERMANY/ALLEMAGNE (DIN)
DIN Deutsches Institut fûr Normung
Burggrafenstrasse 6
Postfach 1107
D-1000 Berlin 30

GREECE/GRÈCE (ELOT)
Hellenic Organization for Standardization
313, Acharnon Street
GR-111 45 Athens

IRELAND/IRLANDE (NSAI)
National Standards Authority of Ireland
Glasnevin
Dublin-9

ITALY/ITALIE (UNI)
Ente Nazionale Italiano di Unificazione
Piazza Armando Diaz 2
1-20123 Milano

NETHERLANDS/PAYS-BAS (NNI)
Nederlands Normalisatie-instituut
Kalfjeslaan 2
P.O. Box 5059
2600 GB Delft

PORTUGAL (IPQ)
Instituto Porugùes da Qualidade
Rua José Estêvão, 83-A
1199 Lisboa Codex

SPAIN/ESPAGNE (AENOR)
Asociación Española de Normalización y
    Certificación
Calle Fernández de la Hoz, 52
28010 Madrid

UNITED KINGDOM/ROYAUME-UNI (BSI)
British Standards Institution
2 Park Street
London W1A 2BS

Note. Luxemburg currently has no standards organisation.

In addition to the EC countries, the EFTA countries shown below also participate:

AUSTRIA/AUTRICHE (ON)
Österreichisches Normungsinstitut
Heinestrasse 38
Postfach 130
A-1021 Wien

FINLAND/FINLANDE (SFS)
Suomen Standardisoimisliitto SFS
P.O. Box 205
SF-00121 Helsinki

NORWAY/NORVÈGE (NSF)
Norges Standardiseringsforbund
Postboks 7020 Homansbyen
N-0306 Oslo 3

SWEDEN/SUÈDE (SIS)
SIS—Standardiseringskommissionen i Sverige
Box 3295
S-103 66 Stockholm

SWITZERLAND/SUISSE (SNV)
Swiss Association for Standardization
Kirchenweg 4
Postfach
CH-8032 Zurich

At the present time the impact and future of CEN standards largely depends on political factors but it is expected that they will be ultimately mandatory for EC member states.

CEN standards will then superçede the equivalent national standards in each member country.

ISO Membership

The EC and EFTA members listed above, together with the following bodies, constitute the total membership of the International Standards Organization.

The letters in parenthesis after the name of the country signify that country's national standards. For example, DIN plugs used on Hi Fi equipment are manufactured to German standards. Plant designed to ANSI standards will be in accordance with American practice.

ALBANIA/ALBANIE (DSMA)
Drejtoria e Standardeve dhe e Mjeteve Matëse në Komisionin e Planit të Shtetit

ALGERIA/ALGÈRIE (INAPI)
Institut algérien de normalisation
et de propriété industrielle

ARGENTINA/ARGENTINE (IRAM)
Instituto Argentino de
Racionalización de Materiales

AUSTRALIA/AUSTRALIE (SAA)
Standards Australia

BANGLADESH (BSTI)
Bangladesh Standards and Testing Institution

BRAZIL/BRÈSIL (ABNT)
Associação Brasileira de Normas Técnicas

BULGARIA/BULGARIE (BDS)
Comité de la qualité

CANADA (SCC)
Standards Council of Canada

CHILE/CHILI (INN)
Instituto Nacional de Normalización

CHINA/CHINE (CSBTS)
China State Bureau of Technical
Supervision

CIS (GOST)
(see RUSSION FEDERATION)

COLOMBIA/COLOMBIE (ICONTEC)
Instituto Colombiano de Normas Técnicas

CUBA (NC)
Comité Estatal de Normalización

CYPRUS/CHYPRE (CYS)
Cyprus Organization for Standards
and Control of Quality

CZECHOSLOVAKIA/TCHECOSLOVAQUIE (CSN)*
Federal Office for Standards and Measurements

EGYPT, Arab Rep. of/
EGYPTE, Rép. arabe d' (EOS)
Egyptian Organization for Standardization and
Quality Control

ETHIOPIA/ÈTHIOPIE (ESA)
Ethiopian Authority for Standardization

GHANA (GSB)
Ghana Standards Board

HUNGARY/HONGRIE (MSZH)
Magyar Szabványügyi Hivatal

INDIA/INDE (BIS)
Bureau of Indian Standards

INDONESIA/INDONÉSIE (DSN)
Dewan Standardisasi Nasional DSN
(Standardization Council of Indonesia)

*Note: this information may change: apply to BSI for latest details.

IRAN, Islamic Rep. of/Rép. islamique d' (ISIRI)
Institute of Standards and Industrial Research of
Iran

IRAQ (COSQC)
Central Organization for Standardization
and Quality Control

ISRAEL/ISRAËL (SII)
Standards Institution of Israel

JAMAICA/JAMAÏQUE (JBS)
Jamaica Bureau of Standards

JAPAN/JAPON (JISC)
Japanese Industrial Standards Committee

KENYA (KEBS)
Kenya Bureau of Standards

**KOREA, Dem. P. Rep. of/
CORÉE, Rép. dém. p. de (CSK)**
Committee for Standardization of the
Democratic People's Republic of Korea

KOREA, Rep. of/CORÉE, Rép. de (KBS)
Bureau of Standards

**LIBYAN ARAB JAMAHIRIYA/
JAMAHIRIYA ARABE LIBYENNE (LNCSM)**
Libyan National Centre for standardization and
Metrology

LITHUANIA (LST)
Lithuanian State Standardization Office

MALAYSIA/MALAISIE (SIRIM)
Standards and Industrial Research Institute
of Malaysia

MEXICO/MEXIQUE (DGN)
Dirección General de Normas

MONGOLIA/MONGOLIE (MNIS)
Mongolian National Institute for Standardization

MOROCCO/MAROC (SNIMA)
Service de normalisation industrielle marocaine

**NEW ZEALAND/NOUVELLE-ZÈLANDE
(SANZ)**
Standards Association of New Zealand

NIGERIA (SON)
Standards Organisation of Nigeria

PAKISTAN (PSI)
Pakistan Standards Institution

**PAPUA NEW GUINEA/
PAPOUASIE-NOUVELLE-GUINÈE (PNGS)**
National Standards Council

PHILIPPINES (BPS)
Bureau of Product Standards

POLAND/POLOGNE (PKNMiJ)
Polish Committee for Standardization, Measures
and Quality Control

ROMANIA/ROUMANIE (IRS)
Institut roumain de normalisation

RUSSIAN FEDERATION (GOST)
State Committee of the Russian Federation
for Standardization, Metrology and Certification

SAUDI ARABIA/ARABIE SAOUDITE (SASO)
Saudi Arabian Standards Organization

SINGAPORE/SINGAPOUR (SISIR)
Singapore Institute of Standards and Industrial
Research

SLOVENIA (SMIS)
Standards and Metrology Institute of Slovenia

**SOUTH AFRICA, Rep. of/
AFRIQUE DU SUD, Rép. d'(SABS)**
South African Bureau of Standards

SRI LANKA (SLSI)
Sri Lanka Standards Institution

SYRIA/SYRIE (SASMO)
Syrian Arab Organization for Standardization and
Metrology

TANZANIA/TANZANIE (TBS)
Tanzania Bureau of Standards

THAILAND/THAÏLANDE (TISI)
Thai Industrial Standards Institute

**TRINIDAD AND TOBAGO/TRINITÉ-ET-
TOBAGO (TTBS)**
Trinidad and Tobago Bureau of Standards

TUNISIA/TUNISIE (INNORPI)
Institut national de la normalisation
et de la propriété industrielle

TURKEY/TURQUIE (TSE)
Türk Standardlari Enstitüsü

USA (ANSI)
American National Standards Institute

USSR/URSS (GOST)
(see RUSSIAN FEDERATION)

VENEZUELA (COVENIN)
Comisión Venezolana de Normas Industriales

VIET NAM, Socialist Republic of/
République socialiste du (TCVN)
General Department for Standardization,
Metrology and Quality

ZIMBABWE (SAZ)
Standards Association of Zimbabwe

Chapter 31
Production drawings

The following three typical drawings are included as examples of draughtsmanship, layout, dimensioning, and tolerancing.

Figs. 31.1 and 31.2 show a pulley and a shaft, and illustrate some aspects of general dimensioning and tolerancing.

Fig. 31.3 shows a partly dimensioned elevation and plan view of a proposed gear-box cover, with a wide application of theoretically exact boxed dimensioning and the associated positional tolerances. To emphasise this style of dimensioning, other dimensions relating to the form of the cover have been omitted.

Completed drawings are usually presented on company standard sheets. Sizes, areas and designations are given at the start of Chapter 5. The layout of a typical sheet contains information required for identification, administration and interpretation and a company could design several types of standard sheets, particularly appropriate to their products.

If drawings are produced manually, then the drawing sheets may be supplied with printed borders and blocks containing relevant notes and headings. With CAD layouts, similar details are stored in the database.

Figs. 31.4 to 31.7, clearly show that documentation can occupy much of the sheet area. The space provided for basic and supplementary information will be decided by the company; its actual position on the sheet may vary with the sheet size.

Fig. 31.1

Fig. 31.2

SECTION X - X

HOLE F
30.025
Ø 30.000

PROPOSED GEAR BOX COVER
MATERIAL L.M.4

Fig. 31.3

Fig. 31.4

Fig. 31.5

Fig. 31.6

COPYRIGHT

This drawing / or document is copyright and the property of Lucas Automotive Ltd. It must not be copied (in whole or in part), used for manufacture or otherwise disclosed without prior written consent of the company. Any copies of this drawing made by any method must also include a copy of this legend.
© Lucas Automotive Ltd 1991

UNLESS OTHERWISE STATED THE FOLLOWING ARE IN ACCORDANCE WITH THE LUCAS STANDARD NO(S) QUOTED
1. SYMBOLS & ABBREVIATIONS TO 51-3-4 & 5
2. I.S.O. METRIC THREADS TO 56-6-4
3. I.S.O. UNIFIED THREADS TO 56-6-2

DIMENSIONS IN MILLIMETRES(mm)
REMOVE SHARP EDGES

UNLESS OTHERWISE STATED TOLERANCES
LINEAR ±   ANGULAR ±
SURFACE ROUGHNESS

ALL DIAMETERS TO BE Ø
UNLESS OTHERWISE STATED

SUPERSEDES	
SIMILAR TO	
REF DRG No	3PX763/155
FIRST USED ON	7204/149

| MATERIAL | SPECIFICATION | 6-18-42 |
| RUBBER | FORM | MOULDING | nl |

HEAT TREATMENT

FINISH

DRAWN R.TRICKETT   CHECKED

TRANSLATED BY   CHECKED

APPROVED   DATE 27/9/91

PROJECTION

CURRENT ☐   SERVICE ☐   CANCELLED ◪

Lucas

Lucas Electronic Unit Injector Systems
Concord Road
London W3 0SE

CLASSIFICATION CODE No.

No. Y7204/140   SHEET 1/1

A3 ENGLISH-

NOTES

nl COMBINED FLASH MUST CONFORM TO BS 6442 : 1984 GRADE N.

EMISSIONS CERTIFIED PRODUCT
THIS PRODUCT HAS BEEN CERTIFICATION TESTED. **NO CHANGE** TO APPROVED PROCESS, MATERIALS OR SUPPLIER IS PERMITTED WITHOUT FORMAL ENGINEERING APPROVAL.

ISSUE No.	DATE & SOURCE	ISSUE No.	DATE & SOURCE	ISSUE No.	DATE & SOURCE
3	7/5/92				
2	21/2/92				
1	7/10/91				

CHANGES

THIS DRAWING MAY ONLY BE ALTERED OR REPLACED IN ACCORDANCE WITH PROCEDURE

DO NOT MEASURE UNDIMENSIONED FEATURES

TITLE SEAL – ACTUATOR BODY

ORIGINAL SCALE   2 : 1

No. Y7204/140

(33.7)

Ø 1.86
Ø 1.70

(23.6)

(R 3.7)

(R 4.7)

(Ø 28.7)

Fig. 31.7

Illustrations of typical layouts are reproduced by kind permission of Lucas Diesel Systems and Lucas Electronic Unit Injector Systems.

An advantage of producing a drawing on a CAD system is that the draughtsman can complete the illustration of the component or assembly on the screen. He may reposition some parts if space permits, to avoid congestion. The outline of the drawing sheet will be taken from the database on a separate layer and placed over the drawing to give a pleasing layout.

Draughtsmen have always derived much personal satisfaction from producing drawings in ink or pencil where the linework and lettering is uniform and neat, and where the illustration completely defines the requirements of the specification. A similar degree of satisfaction can be obtained by producing drawings on CAD equipment. Particular care with the spacings of centre lines, contrasts in linework, cross hatchings, positioning of dimensions, notes, etc, are all small details in their own way, but they collectively contribute to quality, balance and overall clarity. Check the reason for the position of each and every line, ensure that it conforms with the appropriate standard and that it conveys a clear and unambiguous meaning to the reader. Engineering drawings can be considered to be works of art. We both hope you will produce many.

Chapter 32
Drawing solutions

(a)

(b)

(c)

1. Solutions to Fig. 4.17

(a)

(b)

(c)

2. Solutions to Fig. 4.18

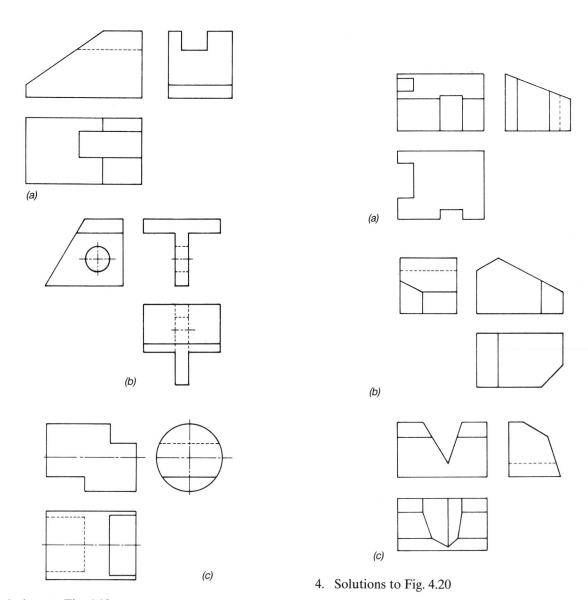

3. Solutions to Fig. 4.19

4. Solutions to Fig. 4.20

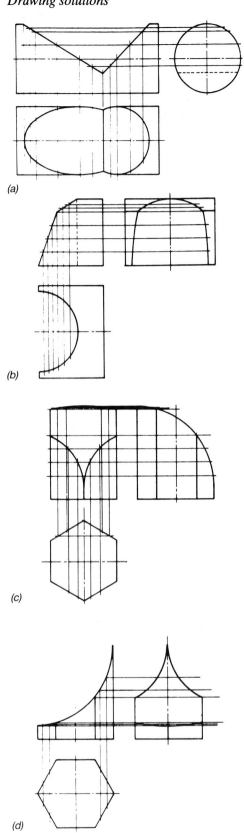

(a)

(b)

(c)

(d)

5.   Solutions to Fig. 4.22

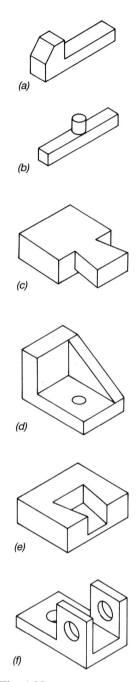

(a)

(b)

(c)

(d)

(e)

(f)

6.   Solutions to Fig. 4.23

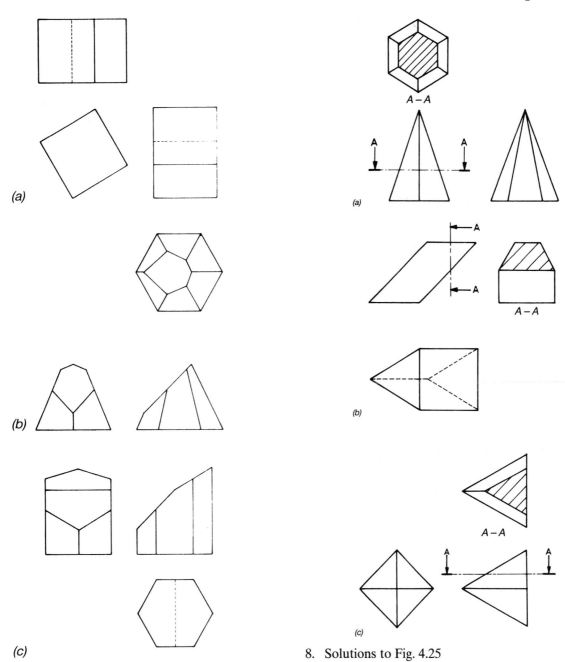

(a)

(b)

(c)

7. Solutions to Fig. 4.24

(a)

(b)

(c)

8. Solutions to Fig. 4.25

(a)

(b)

(c)

(d)

9. Solutions to Fig. 4.26

Index